Dimensions Math®
Teacher's Guide 3B

Authors and Reviewers

Cassandra Turner

Allison Coates

Jenny Kempe

Bill Jackson

Tricia Salerno

Singapore Math Inc.

Published by Singapore Math Inc.

19535 SW 129th Avenue
Tualatin, OR 97062
www.singaporemath.com

Dimensions Math® Teacher's Guide 3B
ISBN 978-1-947226-37-1

First published 2019
Reprinted 2019, 2020

Copyright © 2017 by Singapore Math Inc.
All rights reserved. This book or any portion thereof may not be reproduced or used in any manner whatsoever without the express written permission of the publisher.

Printed in China

Acknowledgments

Editing by the Singapore Math Inc. team.
Design and illustration by Cameron Wray with Carli Fronius.

Contents

Chapter		Lesson	Page
Chapter 8 **Multiplying and Dividing with 6, 7, 8, and 9**		Teaching Notes	1
		Chapter Opener	5
	1	The Multiplication Table of 6	6
	2	The Multiplication Table of 7	10
	3	Multiplying by 6 and 7	14
	4	Dividing by 6 and 7	16
	5	Practice A	18
	6	The Multiplication Table of 8	19
	7	The Multiplication Table of 9	22
	8	Multiplying by 8 and 9	25
	9	Dividing by 8 and 9	27
	10	Practice B	29
		Workbook Pages	31
Chapter 9 **Fractions — Part 1**		Teaching Notes	41
		Chapter Opener	45
	1	Fractions of a Whole	46
	2	Fractions on a Number Line	49
	3	Comparing Fractions with Like Denominators	52
	4	Comparing Fractions with Like Numerators	55
	5	Practice	58
		Workbook Pages	60

Chapter		Lesson	Page
Chapter 10 **Fractions — Part 2**		Teaching Notes	65
		Chapter Opener	69
	1	Equivalent Fractions	70
	2	Finding Equivalent Fractions	72
	3	Simplifying Fractions	75
	4	Comparing Fractions — Part 1	78
	5	Comparing Fractions — Part 2	80
	6	Practice A	83
	7	Adding and Subtracting Fractions — Part 1	85
	8	Adding and Subtracting Fractions — Part 2	88
	9	Practice B	90
		Workbook Pages	92
Chapter 11 **Measurement**		Teaching Notes	101
		Chapter Opener	105
	1	Meters and Centimeters	106
	2	Subtracting from Meters	109
	3	Kilometers	111
	4	Subtracting from Kilometers	113
	5	Liters and Milliliters	115
	6	Kilograms and Grams	118
	7	Word Problems	120
	8	Practice	124
		Review 3	126
		Workbook Pages	129

Chapter		Lesson	Page
Chapter 12 **Geometry**		Teaching Notes	139
		Chapter Opener	145
	1	Circles	146
	2	Angles	149
	3	Right Angles	151
	4	Triangles	154
	5	Properties of Triangles	156
	6	Properties of Quadrilaterals	158
	7	Using a Compass	161
	8	Practice	164
		Workbook Pages	167
Chapter 13 **Area and Perimeter**		Teaching Notes	177
		Chapter Opener	181
	1	Area	182
	2	Units of Area	184
	3	Area of Rectangles	186
	4	Area of Composite Figures	188
	5	Practice A	191
	6	Perimeter	192
	7	Perimeter of Rectangles	194
	8	Area and Perimeter	195
	9	Practice B	197
		Workbook Pages	199

Chapter		Lesson	Page

Chapter 14
Time

		Teaching Notes	209
		Chapter Opener	213
	1	Units of Time	214
	2	Calculating Time — Part 1	218
	3	Practice A	220
	4	Calculating Time — Part 2	221
	5	Calculating Time — Part 3	224
	6	Calculating Time — Part 4	227
	7	Practice B	229
		Workbook Pages	230

Chapter 15
Money

		Teaching Notes	237
		Chapter Opener	241
	1	Dollars and Cents	242
	2	Making $10	245
	3	Adding Money	247
	4	Subtracting Money	250
	5	Word Problems	252
	6	Practice	254
		Review 4	256
		Review 5	258
		Workbook Pages	260

Resources

	Blackline Masters for 3B	269

Dimensions Math® Curriculum

The **Dimensions Math®** series is a Pre-Kindergarten to Grade 5 series based on the pedagogy and methodology of math education in Singapore. The main goal of the **Dimensions Math®** series is to help students develop competence and confidence in mathematics.

The series follows the principles outlined in the Singapore Mathematics Framework below.

Pedagogical Approach and Methodology

- Through Concrete-Pictorial-Abstract development, students view the same concepts over time with increasing levels of abstraction.
- Thoughtful sequencing creates a sense of continuity. The content of each grade level builds on that of preceding grade levels. Similarly, lessons build on previous lessons within each grade.
- Group discussion of solution methods encourages expansive thinking.
- Interesting problems and activities provide varied opportunities to explore and apply skills.
- Hands-on tasks and sharing establish a culture of collaboration.
- Extra practice and extension activities encourage students to persevere through challenging problems.
- Variation in pictorial representation (number bonds, bar models, etc.) and concrete representation (straws, linking cubes, base ten blocks, discs, etc.) broaden student understanding.

Each topic is introduced, then thoughtfully developed through the use of a variety of learning experiences, problem solving, student discourse, and opportunities for mastery of skills. This combination of hands-on practice, in-depth exploration of topics, and mathematical variability in teaching methodology allows students to truly master mathematical concepts.

Singapore Mathematics Framework

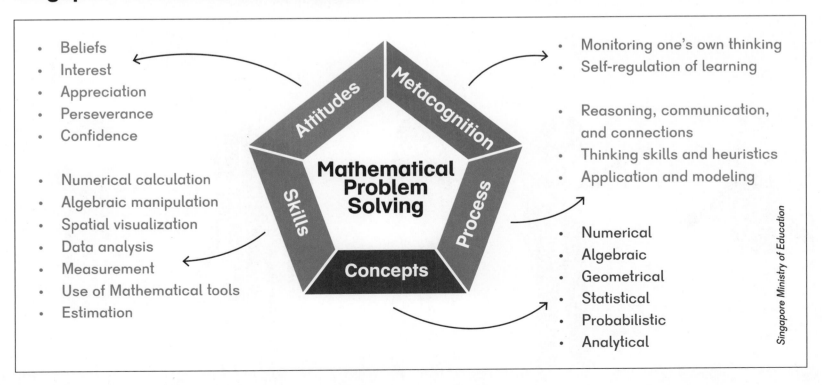

Dimensions Math® Program Materials

Textbooks

Textbooks are designed to help students build a solid foundation in mathematical thinking and efficient problem solving. Careful sequencing of topics, well-chosen problems, and simple graphics foster deep conceptual understanding and confidence. Mental math, problem solving, and correct computation are given balanced attention in all grades. As skills are mastered, students move to increasingly sophisticated concepts within and across grade levels.

Students work through the textbook lessons with the help of five friends: Emma, Alex, Sofia, Dion, and Mei. The characters appear throughout the series and help students develop metacognitive reasoning through questions, hints, and ideas.

A pencil icon ▶ at the end of the textbook lessons links to exercises in the workbooks.

Workbooks

Workbooks provide additional problems that range from basic to challenging. These allow students to independently review and practice the skills they have learned.

Teacher's Guides

Teacher's Guides include lesson plans, mathematical background, games, helpful suggestions, and comprehensive resources for daily lessons.

Tests

Tests contain differentiated assessments to systematically evaluate student progress.

Emma   Alex   Sofia   Dion   Mei

Online Resources

The following can be downloaded from dimensionsmath.com.

- **Blackline Masters** used for various hands-on tasks.

- **Material Lists** for each chapter and lesson, so teachers and classroom helpers can prepare ahead of time.

- **Activities** that can done with students who need more practice or a greater challenge, organized by concept, chapter, and lesson.

- **Standards Alignments** for various states.

Using the Teacher's Guide

This guide is designed to assist in planning daily lessons. It should be considered a helping hand between the curriculum and the classroom. It provides introductory notes on mathematical content, key points, and suggestions for activities. It also includes ideas for differentiation within each lesson, and answers and solutions to textbook and workbook problems.

Each chapter of the guide begins with the following.

- **Overview**

  Includes objectives and suggested number of class periods for each chapter.

- **Notes**

  Highlights key learning points, provides background on math concepts, explains the purpose of certain activities, and helps teachers understand the flow of topics throughout the year.

- **Materials**

  Lists materials, manipulatives, and Blackline Masters used in the Think and Learn sections of the guide. It also includes suggested storybooks. Many common classroom manipulatives are used throughout the curriculum. When a lesson refers to a whiteboard and markers, any writing materials can be used. Blackline Masters can be found at dimensionsmath.com.

The guide goes through the Chapter Openers, Daily Lessons, and Practices of each chapter, and cumulative reviews in the following general format.

- **Chapter Opener**

  Provides talking points for discussion to prepare students for the math concepts to be introduced.

- **Think**

  Offers structure for teachers to guide student inquiry. Provides various methods and activities to solve initial textbook problems or tasks.

- **Learn**

  Guides teachers to analyze student methods from Think to arrive at the main concepts of the lesson through discussion and study of the pictorial representations in the textbook.

- **Do**

  Expands on specific problems with strategies, additional practice, and remediation.

● Activities

Allows students to practice concepts through individual, small group, and whole group hands-on tasks and games, including suggestions for outdoor play (most of which can be modified for a gymnasium or classroom).

Level of difficulty in the games and activities are denoted by the following symbols.

- ● Foundational activities
- ▲ On-level activities
- ★ Challenge or extension activities

● Brain Works

Provides opportunities for students to extend their mathematical thinking.

Discussion is a critical component of each lesson. Teachers are encouraged to let students discuss their reasoning. As each classroom is different, this guide does not anticipate all situations. The following questions can help students articulate their thinking and increase their mastery:

- Why? How do you know?
- Can you explain that?
- Can you draw a picture of that?
- Is your answer reasonable? How do you know?
- How is this task like the one we did before? How is it different?
- What is alike and what is different about…?
- Can you solve that a different way?
- Yes! You're right! How do you know it's true?
- What did you learn before that can help you solve this problem?
- Can you summarize what your classmate shared?
- What conclusion can you draw from the data?

Each lesson is designed to take one day. If your calendar allows, you may choose to spend more than one day on certain lessons. Throughout the guide, there are notes to extend on learning activities to make them more challenging. Lesson structures and activities do not have to conform exactly to what is shown in the guide. Teachers are encouraged to exercise their discretion in using this material in a way that best suits their classes.

Textbooks are designed to last multiple years. Textbook problems with a ▇ (or a blank line for terms) are meant to invite active participation.

Dimensions Math® Scope & Sequence

PKA

Chapter 1
Match, Sort, and Classify

Red and Blue
Yellow and Green
Color Review
Soft and Hard
Rough, Bumpy, and Smooth
Sticky and Grainy
Size — Part 1
Size — Part 2
Sort Into Two Groups
Practice

Chapter 2
Compare Objects

Big and Small
Long and Short
Tall and Short
Heavy and Light
Practice

Chapter 3
Patterns

Movement Patterns
Sound Patterns
Create Patterns
Practice

Chapter 4
Numbers to 5 — Part 1

Count 1 to 5 — Part 1
Count 1 to 5 — Part 2
Count Back
Count On and Back
Count 1 Object
Count 2 Objects
Count Up to 3 Objects
Count Up to 4 Objects
Count Up to 5 Objects
How Many? — Part 1
How Many? — Part 2
How Many Now? — Part 1
How Many Now? — Part 2
Practice

Chapter 5
Numbers to 5 — Part 2

1, 2, 3
1, 2, 3, 4, 5 — Part 1
1, 2, 3, 4, 5 — Part 2
How Many? — Part 1
How Many? — Part 2
How Many Do You See?
How Many Do You See Now?
Practice

Chapter 6
Numbers to 10 — Part 1

0
Count to 10 — Part 1
Count to 10 — Part 2
Count Back
Order Numbers
Count Up to 6 Objects
Count Up to 7 Objects
Count Up to 8 Objects
Count Up to 9 Objects
Count Up to 10 Objects — Part 1
Count Up to 10 Objects — Part 2
How Many?
Practice

Chapter 7
Numbers to 10 — Part 2

6
7
8
9
10
0 to 10
Count and Match — Part 1
Count and Match — Part 2
Practice

PKB

Chapter 8
Ordinal Numbers

First
Second and Third
Fourth and Fifth
Practice

Chapter 9
Shapes and Solids

Cubes, Cylinders, and Spheres
Cubes
Positions
Build with Solids
Rectangles and Circles
Squares
Triangles

Squares, Circles, Rectangles, and Triangles — Part 1
Squares, Circles, Rectangles, and Triangles — Part 2
Practice

Chapter 10
Compare Sets

Match Objects
Which Set Has More?
Which Set Has Fewer?
More or Fewer?
Practice

Chapter 11
Compose and Decompose

Altogether — Part 1
Altogether — Part 2
Show Me
What's the Other Part? — Part 1
What's the Other Part? — Part 2
Practice

Chapter 12
Explore Addition and Subtraction

Add to 5 — Part 1
Add to 5 — Part 2
Two Parts Make a Whole
How Many in All?
Subtract Within 5 — Part 1
Subtract Within 5 — Part 2
How Many Are Left?

Practice

Chapter 13
Cumulative Review

Review 1 Match and Color
Review 2 Big and Small
Review 3 Heavy and Light
Review 4 Count to 5
Review 5 Count 5 Objects
Review 6 0
Review 7 Count Beads
Review 8 Patterns
Review 9 Length
Review 10 How Many?
Review 11 Ordinal Numbers
Review 12 Solids and Shapes
Review 13 Which Set Has More?
Review 14 Which Set Has Fewer?
Review 15 Put Together
Review 16 Subtraction
Looking Ahead 1 Sequencing — Part 1
Looking Ahead 2 Sequencing — Part 2
Looking Ahead 3 Categorizing
Looking Ahead 4 Addition
Looking Ahead 5 Subtraction
Looking Ahead 6 Getting Ready to Write Numerals
Looking Ahead 7 Reading and Math

KA

Chapter 1
Match, Sort, and Classify

Left and Right
Same and Similar
Look for One That Is Different
How Does it Feel?
Match the Things That Go Together
Sort
Practice

Chapter 2
Numbers to 5

Count to 5
Count Things Up to 5
Recognize the Numbers 1 to 3
Recognize the Numbers 4 and 5
Count and Match
Write the Numbers 1 and 2
Write the Number 3
Write the Number 4
Trace and Write 1 to 5
Zero
Picture Graphs
Practice

Chapter 3
Numbers to 10

Count 1 to 10
Count Up to 7 Things
Count Up to 9 Things
Count Up to 10 Things — Part 1

Dimensions Math® Scope & Sequence

Count Up to 10 Things — Part 2
Recognize the Numbers 6 to 10
Write the Numbers 6 and 7
Write the Numbers 8, 9, and 10
Write the Numbers 6 to 10
Count and Write the Numbers 1 to 10
Ordinal Positions
One More Than
Practice

Chapter 4
Shapes and Solids

Curved or Flat
Solid Shapes
Closed Shapes
Rectangles
Squares
Circles and Triangles
Where is It?
Hexagons
Sizes and Shapes
Combine Shapes
Graphs
Practice

Chapter 5
Compare Height, Length, Weight, and Capacity

Comparing Height
Comparing Length
Height and Length — Part 1
Height and Length — Part 2
Weight — Part 1

Weight — Part 2
Weight — Part 3
Capacity — Part 1
Capacity — Part 2
Practice

Chapter 6
Comparing Numbers Within 10

Same and More
More and Fewer
More and Less
Practice — Part 1
Practice — Part 2

KB

Chapter 7
Numbers to 20

Ten and Some More
Count Ten and Some More
Two Ways to Count
Numbers 16 to 20
Number Words 0 to 10
Number Words 11 to 15
Number Words 16 to 20
Number Order
1 More Than or Less Than
Practice — Part 1
Practice — Part 2

Chapter 8
Number Bonds

Putting Numbers Together — Part 1

Putting Numbers Together — Part 2
Parts Making a Whole
Look for a Part
Number Bonds for 2, 3, and 4
Number Bonds for 5
Number Bonds for 6
Number Bonds for 7
Number Bonds for 8
Number Bonds for 9
Number Bonds for 10
Practice — Part 1
Practice — Part 2
Practice — Part 3

Chapter 9
Addition

Introduction to Addition — Part 1
Introduction to Addition — Part 2
Introduction to Addition — Part 3
Addition
Count On — Part 1
Count On — Part 2
Add Up to 3 and 4
Add Up to 5 and 6
Add Up to 7 and 8
Add Up to 9 and 10
Addition Practice
Practice

Chapter 10
Subtraction

Take Away to Subtract — Part 1

Take Away to Subtract — Part 2
Take Away to Subtract — Part 3
Take Apart to Subtract — Part 1
Take Apart to Subtract — Part 2
Count Back
Subtract Within 5
Subtract Within 10 — Part 1
Subtract Within 10 — Part 2
Practice

Chapter 11
Addition and Subtraction

Add and Subtract
Practice Addition and Subtraction
Part-Whole Addition and Subtraction
Add to or Take Away
Put Together or Take Apart
Practice

Chapter 12
Numbers to 100

Count by Tens — Part 1
Count by Tens — Part 2
Numbers to 30
Numbers to 40
Numbers to 50
Numbers to 80
Numbers to 100 — Part 1
Numbers to 100 — Part 2
Count by Fives — Part 1
Count by Fives — Part 2

Practice

Chapter 13
Time

Day and Night
Learning About the Clock
Telling Time to the Hour — Part 1
Telling Time to the Hour — Part 2
Practice

Chapter 14
Money

Coins
Pennies
Nickels
Dimes
Quarters
Practice

1A

Chapter 1
Numbers to 10

Numbers to 10
The Number 0
Order Numbers
Compare Numbers
Practice

Chapter 2
Number Bonds

Make 6
Make 7
Make 8

Make 9
Make 10 — Part 1
Make 10 — Part 2
Practice

Chapter 3
Addition

Addition as Putting Together
Addition as Adding More
Addition with 0
Addition with Number Bonds
Addition by Counting On
Make Addition Stories
Addition Facts
Practice

Chapter 4
Subtraction

Subtraction as Taking Away
Subtraction as Taking Apart
Subtraction by Counting Back
Subtraction with 0
Make Subtraction Stories
Subtraction with Number Bonds
Addition and Subtraction
Make Addition and Subtraction Story Problems
Subtraction Facts
Practice
Review 1

Chapter 5
Numbers to 20

Numbers to 20
Add or Subtract Tens or Ones
Order Numbers to 20

Dimensions Math® Scope & Sequence

Compare Numbers to 20
Addition
Subtraction
Practice

Chapter 6
Addition to 20

Add by Making 10 — Part 1
Add by Making 10 — Part 2
Add by Making 10 — Part 3
Addition Facts to 20
Practice

Chapter 7
Subtraction Within 20

Subtract from 10 — Part 1
Subtract from 10 — Part 2
Subtract the Ones First
Word Problems
Subtraction Facts Within 20
Practice

Chapter 8
Shapes

Solid and Flat Shapes
Grouping Shapes
Making Shapes
Practice

Chapter 9
Ordinal Numbers

Naming Positions
Word Problems
Practice
Review 2

1B

Chapter 10
Length

Comparing Lengths Directly
Comparing Lengths Indirectly
Comparing Lengths with Units
Practice

Chapter 11
Comparing

Subtraction as Comparison
Making Comparison Subtraction Stories
Picture Graphs
Practice

Chapter 12
Numbers to 40

Numbers to 40
Tens and Ones
Counting by Tens and Ones
Comparing
Practice

Chapter 13
Addition and Subtraction Within 40

Add Ones
Subtract Ones
Make the Next Ten
Use Addition Facts
Subtract from Tens
Use Subtraction Facts
Add Three Numbers
Practice

Chapter 14
Grouping and Sharing

Adding Equal Groups
Sharing
Grouping
Practice

Chapter 15
Fractions

Halves
Fourths
Practice
Review 3

Chapter 16
Numbers to 100

Numbers to 100
Tens and Ones
Count by Ones or Tens
Compare Numbers to 100
Practice

Chapter 17
Addition and Subtraction Within 100

Add Ones — Part 1
Add Tens
Add Ones — Part 2
Add Tens and Ones — Part 1
Add Tens and Ones — Part 2
Subtract Ones — Part 1
Subtract from Tens
Subtract Ones — Part 2
Subtract Tens

Subtract Tens and Ones — Part 1
Subtract Tens and Ones — Part 2
Practice

Chapter 18
Time

Telling Time to the Hour
Telling Time to the Half Hour
Telling Time to the 5 Minutes
Practice

Chapter 19
Money

Coins
Counting Money
Bills
Shopping
Practice
Review 4

2A

Chapter 1
Numbers to 1,000

Tens and Ones
Counting by Tens or Ones
Comparing Tens and Ones
Hundreds, Tens, and Ones
Place Value
Comparing Hundreds, Tens, and Ones
Counting by Hundreds, Tens, or Ones
Practice

Chapter 2
Addition and Subtraction — Part 1

Strategies for Addition
Strategies for Subtraction
Parts and Whole
Comparison
Practice

Chapter 3
Addition and Subtraction — Part 2

Addition Without Regrouping
Subtraction Without Regrouping
Addition with Regrouping Ones
Addition with Regrouping Tens
Addition with Regrouping Tens and Ones
Practice A
Subtraction with Regrouping from Tens
Subtraction with Regrouping from Hundreds
Subtraction with Regrouping from Two Places
Subtraction with Regrouping across Zeros
Practice B
Practice C

Chapter 4
Length

Centimeters
Estimating Length in Centimeters
Meters
Estimating Length in Meters
Inches
Using Rulers
Feet
Practice

Chapter 5
Weight

Grams
Kilograms
Pounds
Practice
Review 1

Chapter 6
Multiplication and Division

Multiplication — Part 1
Multiplication — Part 2
Practice A
Division — Part 1
Division — Part 2
Multiplication and Division
Practice B

Chapter 7
Multiplication and Division of 2, 5, and 10

The Multiplication Table of 5
Multiplication Facts of 5
Practice A
The Multiplication Table of 2
Multiplication Facts of 2
Practice B
The Multiplication Table of 10
Dividing by 2

Dimensions Math® Scope & Sequence

Dividing by 5 and 10
Practice C
Word Problems
Review 2

2B

Chapter 8
Mental Calculation

Adding Ones Mentally
Adding Tens Mentally
Making 100
Adding 97, 98, or 99
Practice A
Subtracting Ones Mentally
Subtracting Tens Mentally
Subtracting 97, 98, or 99
Practice B
Practice C

Chapter 9
Multiplication and Division of 3 and 4

The Multiplication Table of 3
Multiplication Facts of 3
Dividing by 3
Practice A
The Multiplication Table of 4
Multiplication Facts of 4
Dividing by 4
Practice B
Practice C

Chapter 10
Money

Making $1
Dollars and Cents
Making Change
Comparing Money
Practice A
Adding Money
Subtracting Money
Practice B

Chapter 11
Fractions

Halves and Fourths
Writing Unit Fractions
Writing Fractions
Fractions that Make 1 Whole
Comparing and Ordering Fractions
Practice
Review 3

Chapter 12
Time

Telling Time
Time Intervals
A.M. and P.M.
Practice

Chapter 13
Capacity

Comparing Capacity
Units of Capacity
Practice

Chapter 14
Graphs

Picture Graphs
Bar Graphs
Practice

Chapter 15
Shapes

Straight and Curved Sides
Polygons
Semicircles and Quarter-circles
Patterns
Solid Shapes
Practice
Review 4
Review 5

3A

Chapter 1
Numbers to 10,000

Numbers to 10,000
Place Value — Part 1
Place Value — Part 2
Comparing Numbers
The Number Line
Practice A
Number Patterns
Rounding to the Nearest Thousand
Rounding to the Nearest Hundred
Rounding to the Nearest Ten
Practice B

Chapter 2
Addition and Subtraction — Part 1

Mental Addition — Part 1
Mental Addition — Part 2
Mental Subtraction — Part 1
Mental Subtraction — Part 2
Making 100 and 1,000
Strategies for Numbers Close to Hundreds
Practice A
Sum and Difference
Word Problems — Part 1
Word Problems — Part 2
2-Step Word Problems
Practice B

Chapter 3
Addition and Subtraction — Part 2

Addition with Regrouping
Subtraction with Regrouping — Part 1
Subtraction with Regrouping — Part 2
Estimating Sums and Differences — Part 1
Estimating Sums and Differences — Part 2
Word Problems
Practice

Chapter 4
Multiplication and Division

Looking Back at Multiplication
Strategies for Finding the Product
Looking Back at Division
Multiplying and Dividing with 0 and 1
Division with Remainders
Odd and Even Numbers
Word Problems — Part 1
Word Problems — Part 2
2-Step Word Problems
Practice
Review 1

Chapter 5
Multiplication

Multiplying Ones, Tens, and Hundreds
Multiplication Without Regrouping
Multiplication with Regrouping Tens
Multiplication with Regrouping Ones
Multiplication with Regrouping Ones and Tens
Practice A
Multiplying a 3-Digit Number with Regrouping Once
Multiplication with Regrouping More Than Once
Practice B

Chapter 6
Division

Dividing Tens and Hundreds
Dividing a 2-Digit Number by 2 — Part 1
Dividing a 2-Digit Number by 2 — Part 2
Dividing a 2-Digit Number by 3, 4, and 5
Practice A
Dividing a 3-Digit Number by 2
Dividing a 3-Digit Number by 3, 4, and 5
Dividing a 3-Digit Number, Quotient is 2 Digits
Practice B

Chapter 7
Graphs and Tables

Picture Graphs and Bar Graphs
Bar Graphs and Tables
Practice
Review 2

3B

Chapter 8
Multiplying and Dividing with 6, 7, 8, and 9

The Multiplication Table of 6
The Multiplication Table of 7
Multiplying by 6 and 7
Dividing by 6 and 7
Practice A
The Multiplication Table of 8

Dimensions Math® Scope & Sequence

The Multiplication Table of 9
Multiplying by 8 and 9
Dividing by 8 and 9
Practice B

Chapter 9
Fractions — Part 1

Fractions of a Whole
Fractions on a Number Line
Comparing Fractions with Like Denominators
Comparing Fractions with Like Numerators
Practice

Chapter 10
Fractions — Part 2

Equivalent Fractions
Finding Equivalent Fractions
Simplifying Fractions
Comparing Fractions — Part 1
Comparing Fractions — Part 2
Practice A
Adding and Subtracting Fractions — Part 1
Adding and Subtracting Fractions — Part 2
Practice B

Chapter 11
Measurement

Meters and Centimeters
Subtracting from Meters
Kilometers
Subtracting from Kilometers
Liters and Milliliters
Kilograms and Grams

Word Problems
Practice
Review 3

Chapter 12
Geometry

Circles
Angles
Right Angles
Triangles
Properties of Triangles
Properties of Quadrilaterals
Using a Compass
Practice

Chapter 13
Area and Perimeter

Area
Units of Area
Area of Rectangles
Area of Composite Figures
Practice A
Perimeter
Perimeter of Rectangles
Area and Perimeter
Practice B

Chapter 14
Time

Units of Time
Calculating Time — Part 1
Practice A
Calculating Time — Part 2
Calculating Time — Part 3
Calculating Time — Part 4
Practice B

Chapter 15
Money

Dollars and Cents
Making $10
Adding Money
Subtracting Money
Word Problems
Practice
Review 4
Review 5

4A

Chapter 1
Numbers to One Million

Numbers to 100,000
Numbers to 1,000,000
Number Patterns
Comparing and Ordering Numbers
Rounding 5-Digit Numbers
Rounding 6-Digit Numbers
Calculations and Place Value
Practice

Chapter 2
Addition and Subtraction

Addition
Subtraction
Other Ways to Add and Subtract — Part 1
Other Ways to Add and Subtract — Part 2
Word Problems

Practice

Chapter 3
Multiples and Factors

Multiples
Common Multiples
Factors
Prime Numbers and
    Composite Numbers
Common Factors
Practice

Chapter 4
Multiplication

Mental Math for Multiplication
Multiplying by a 1-Digit
    Number — Part 1
Multiplying by a 1-Digit
    Number — Part 2
Practice A
Multiplying by a Multiple of 10
Multiplying by a 2-Digit
    Number — Part 1
Multiplying by a 2-Digit
    Number — Part 2
Practice B

Chapter 5
Division

Mental Math for Division
Estimation and Division
Dividing 4-Digit Numbers
Practice A
Word Problems
Challenging Word Problems
Practice B
Review 1

Chapter 6
Fractions

Equivalent Fractions
Comparing and Ordering
    Fractions
Improper Fractions and Mixed
    Numbers
Practice A
Expressing an Improper
    Fraction as a Mixed
    Number
Expressing a Mixed Number
    as an Improper Fraction
Fractions and Division
Practice B

Chapter 7
Adding and Subtracting Fractions

Adding and Subtracting
    Fractions — Part 1
Adding and Subtracting
    Fractions — Part 2
Adding a Mixed Number and
    a Fraction
Adding Mixed Numbers
Subtracting a Fraction from
    a Mixed Number
Subtracting Mixed Numbers
Practice

Chapter 8
Multiplying a Fraction and a Whole Number

Multiplying a Unit Fraction
    by a Whole Number

Multiplying a Fraction by a
    Whole Number — Part 1
Multiplying a Fraction by a
    Whole Number — Part 2
Fraction of a Set
Multiplying a Whole Number
    by a Fraction — Part 1
Multiplying a Whole Number
    by a Fraction — Part 2
Word Problems — Part 1
Word Problems — Part 2
Practice

Chapter 9
Line Graphs and Line Plots

Line Graphs
Drawing Line Graphs
Line Plots
Practice
Review 2

4B

Chapter 10
Measurement

Metric Units of Measurement
Customary Units of Length
Customary Units of Weight
Customary Units of Capacity
Units of Time
Practice A
Fractions and Measurement
    — Part 1
Fractions and Measurement
    — Part 2
Practice B

Dimensions Math® Scope & Sequence

Chapter 11
Area and Perimeter

Area of Rectangles — Part 1
Area of Rectangles — Part 2
Area of Composite Figures
Perimeter — Part 1
Perimeter — Part 2
Practice

Chapter 12
Decimals

Tenths — Part 1
Tenths — Part 2
Hundredths — Part 1
Hundredths — Part 2
Expressing Decimals as Fractions in Simplest Form
Expressing Fractions as Decimals
Practice A
Comparing and Ordering Decimals
Rounding Decimals
Practice B

Chapter 13
Addition and Subtraction of Decimals

Adding and Subtracting Tenths
Adding Tenths with Regrouping
Subtracting Tenths with Regrouping
Practice A
Adding Hundredths
Subtracting from 1 and 0.1
Subtracting Hundredths
Money, Decimals, and Fractions

Practice B
Review 3

Chapter 14
Multiplication and Division of Decimals

Multiplying Tenths and Hundredths
Multiplying Decimals by a Whole Number — Part 1
Multiplying Decimals by a Whole Number — Part 2
Practice A
Dividing Tenths and Hundredths
Dividing Decimals by a Whole Number — Part 1
Dividing Decimals by a Whole Number — Part 2
Dividing Decimals by a Whole Number — Part 3
Practice B

Chapter 15
Angles

The Size of Angles
Measuring Angles
Drawing Angles
Adding and Subtracting Angles
Reflex Angles
Practice

Chapter 16
Lines and Shapes

Perpendicular Lines
Parallel Lines
Drawing Perpendicular and Parallel Lines
Quadrilaterals

Lines of Symmetry
Symmetrical Figures and Patterns
Practice

Chapter 17
Properties of Cuboids

Cuboids
Nets of Cuboids
Faces and Edges of Cuboids
Practice
Review 4
Review 5

5A

Chapter 1
Whole Numbers

Numbers to One Billion
Multiplying by 10, 100, and 1,000
Dividing by 10, 100, and 1,000
Multiplying by Tens, Hundreds, and Thousands
Dividing by Tens, Hundreds, and Thousands
Practice

Chapter 2
Writing and Evaluating Expressions

Expressions with Parentheses
Order of Operations — Part 1
Order of Operations — Part 2

Other Ways to Write and Evaluate Expressions
Word Problems — Part 1
Word Problems — Part 2
Practice

Chapter 3
Multiplication and Division

Multiplying by a 2-digit Number — Part 1
Multiplying by a 2-digit Number — Part 2
Practice A
Dividing by a Multiple of Ten
Divide a 2-digit Number by a 2-digit Number
Divide a 3-digit Number by a 2-digit Number — Part 1
Divide a 3-digit Number by a 2-digit Number — Part 2
Divide a 4-digit Number by a 2-digit Number
Practice B

Chapter 4
Addition and Subtraction of Fractions

Fractions and Division
Adding Unlike Fractions
Subtracting Unlike Fractions
Practice A
Adding Mixed Numbers — Part 1
Adding Mixed Numbers — Part 2
Subtracting Mixed Numbers — Part 1
Subtracting Mixed Numbers — Part 2
Practice B
Review 1

Chapter 5
Multiplication of Fractions

Multiplying a Fraction by a Whole Number
Multiplying a Whole Number by a Fraction
Word Problems — Part 1
Practice A
Multiplying a Fraction by a Unit Fraction
Multiplying a Fraction by a Fraction — Part 1
Multiplying a Fraction by a Fraction — Part 2
Multiplying Mixed Numbers
Word Problems — Part 2
Fractions and Reciprocals
Practice B

Chapter 6
Division of Fractions

Dividing a Unit Fraction by a Whole Number
Dividing a Fraction by a Whole Number
Practice A
Dividing a Whole Number by a Unit Fraction
Dividing a Whole Number by a Fraction
Word Problems
Practice B

Chapter 7
Measurement

Fractions and Measurement Conversions
Fractions and Area
Practice A
Area of a Triangle — Part 1
Area of a Triangle — Part 2
Area of Complex Figures
Practice B

Chapter 8
Volume of Solid Figures

Cubic Units
Volume of Cuboids
Finding the Length of an Edge
Practice A
Volume of Complex Shapes
Volume and Capacity — Part 1
Volume and Capacity — Part 2
Practice B
Review 2

5B

Chapter 9
Decimals

Thousandths
Place Value to Thousandths
Comparing Decimals
Rounding Decimals
Practice A
Multiply Decimals by 10, 100, and 1,000
Divide Decimals by 10, 100, and 1,000

Dimensions Math® Scope & Sequence

Conversion of Measures
Mental Calculation
Practice B

Chapter 10
The Four Operations of Decimals

Adding Decimals to Thousandths
Subtracting Decimals
Multiplying by 0.1 or 0.01
Multiplying by a Decimal
Practice A
Dividing by a Whole Number — Part 1
Dividing by a Whole Number — Part 2
Dividing a Whole Number by 0.1 and 0.01
Dividing a Whole Number by a Decimal
Practice B

Chapter 11
Geometry

Measuring Angles
Angles and Lines
Classifying Triangles
The Sum of the Angles in a Triangle
The Exterior Angle of a Triangle
Classifying Quadrilaterals
Angles of Quadrilaterals — Part 1
Angles of Quadrilaterals — Part 2

Drawing Triangles and Quadrilaterals
Practice

Chapter 12
Data Analysis and Graphs

Average — Part 1
Average — Part 2
Line Plots
Coordinate Graphs
Straight Line Graphs
Practice
Review 3

Chapter 13
Ratio

Finding the Ratio
Equivalent Ratios
Finding a Quantity
Comparing Three Quantities
Word Problems
Practice

Chapter 14
Rate

Finding the Rate
Rate Problems — Part 1
Rate Problems — Part 2
Word Problems
Practice

Chapter 15
Percentage

Meaning of Percentage
Expressing Percentages as Fractions

Percentages and Decimals
Expressing Fractions as Percentages
Practice A
Percentage of a Quantity
Word Problems
Practice B
Review 4
Review 5

Chapter 8 Multiplying and Dividing with 6, 7, 8, and 9

Overview

Suggested number of class periods: 10–11

Lesson		Page	Resources		Objectives
	Chapter Opener	p. 5	TB:	p. 1	Investigate multiplication and division of 6, 7, 8, and 9.
1	The Multiplication Table of 6	p. 6	TB: WB:	p. 2 p. 1	Use the properties of operations to derive new facts from known facts in the multiplication table of 6.
2	The Multiplication Table of 7	p. 10	TB: WB:	p. 7 p. 4	Use the properties of operations to derive new facts from known facts in the multiplication table of 7.
3	Multiplying by 6 and 7	p. 14	TB: WB:	p. 12 p. 7	Multiply a number of up to 3 digits by 6 or 7.
4	Dividing by 6 and 7	p. 16	TB: WB:	p. 14 p. 10	Divide a number of up to 3 digits by 6 and 7.
5	Practice A	p. 18	TB: WB:	p. 17 p. 13	Practice multiplying and dividing by 6 and 7. Solve word problems involving multiplication and division.
6	The Multiplication Table of 8	p. 19	TB: WB:	p. 18 p. 17	Use the properties of operations to derive new facts from known facts in the multiplication table of 8.
7	The Multiplication Table of 9	p. 22	TB: WB:	p. 23 p. 20	Use the properties of operations to derive new facts from known facts in the multiplication table of 9.
8	Multiplying by 8 and 9	p. 25	TB: WB:	p. 27 p. 23	Multiply a number of up to 3 digits by 8 or 9.
9	Dividing by 8 and 9	p. 27	TB: WB:	p. 29 p. 26	Divide a number of up to 3 digits by 8 and 9.
10	Practice B	p. 29	TB: WB:	p. 31 p. 29	Practice multiplying and dividing by 8 and 9. Solve word problems involving multiplication and division.
	Workbook Solutions	p. 31			

Chapter 8 Multiplying and Dividing with 6, 7, 8, and 9

Notes

In this chapter, students learn the remaining multiplication and division facts for 6, 7, 8, and 9. In Dimensions Math 2, students learned the multiplication and division facts for 2, 3, 4, 5, and 10, which were then reviewed in Dimensions Math 3A. In Dimensions Math 3A, they learned how to multiply or divide a number of up to 3 digits by 2, 3, 4, and 5.

In this chapter, students will continue to use facts they know to find facts they have not yet learned. They already know many of the facts for 6, 7, 8, and 9, through an understanding of the commutative property of multiplication: 4 × 7 = 7 × 4.

Students will also continue to informally apply the distributive property, as they did in Dimensions Math 3A Chapter 4 Lesson 2: Strategies for Finding the Product. For example, 6 × 7 can be thought of as the sum of the products of 6 × 5 and 6 × 2.

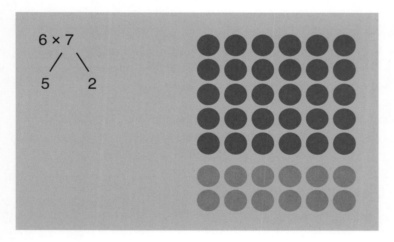

From the diagram, students can see that they can add the products of 6 × 5 and 6 × 2 to get the product of 6 × 7.

Students do not need to learn the names of the properties (commutative and distributive) or rules about order of operations at this point, nor will they need to write out an equation such as 6 × 7 = 6 × (5 + 2) = 6 × 5 + 6 × 2.

On a multiplication chart, there are 16 facts left to learn. However, as students already understand that they can multiply in any order, there are really only 10 new facts in this chapter:

6 × 6	6 × 7	6 × 8	6 × 9
	7 × 7	7 × 8	7 × 9
		8 × 8	8 × 9
			9 × 9

×	1	2	3	4	5	6	7	8	9	10
1	1	2	3	4	5	6	7	8	9	10
2	2	4	6	8	10	12	14	16	18	20
3	3	6	9	12	15	18	21	24	27	30
4	4	8	12	16	20	24	28	32	36	40
5	5	10	15	20	25	30	35	40	45	50
6	6	12	18	24	30					60
7	7	14	21	28	35					70
8	8	16	24	32	40					80
9	9	18	27	36	45					90
10	10	20	30	40	50	60	70	80	90	100

Students will apply learned facts for 6, 7, 8, and 9 to the multiplication and division algorithms. These skills will help students master the facts while they continue to practice algorithms.

Students' grasp of the multiplication and division algorithms should be beyond the need for concrete manipulatives. Using place-value discs for these facts can be cumbersome as many discs are needed for groups of 7, 8, or 9. Therefore, if students continue to struggle with the steps in an algorithm, provide additional practice using place-value discs with multipliers or divisors of 2, 3, or 4.

In Dimensions Math 3A Chapter 6: Division, the division algorithm was introduced and practiced with problems that involved sharing. Sharing (or partitive) division begins with the total number of objects and puts them into a fixed number of groups to find out how many there are in each group.

Chapter 8 Multiplying and Dividing with 6, 7, 8, and 9

For example:

- 246 jelly beans are shared between 2 friends. How many jelly beans does each friend receive?
  246 ÷ 2 = 128 jelly beans

In Dimensions Math 3A Chapter 6 Lesson 4: Dividing a 2-Digit Number by 3, 4, and 5, students also learned that they can use the same division algorithm for grouping (or quotative) situations.

Division as grouping begins with the total number of objects and puts them into equal groups of a given amount to find out how many groups we can make.

For example:

- There are 246 jelly beans. Each friend receives 2 jelly beans. How many friends will get jelly beans? 246 ÷ 2 = 128 friends

While place-value discs do not illustrate grouping situations, the long division algorithm works exactly the same for both grouping and sharing situations. Students should know by now that the answer is the same, whether the situation involves sharing or grouping. Note that students use the terms "sharing" and "grouping," not "partitive division" and "quotative division."

Chapter 8 Multiplying and Dividing with 6, 7, 8, and 9 — Materials

Materials

- Blank paper
- Index cards
- Paper plates with the center cut out
- Place-value discs
- Whiteboards

Blackline Masters

- Array Dot Cards — 6
- Array Dot Cards — 7
- Kaboom Cards
- Multiplication and Division Fact Cards for 6
- Multiplication and Division Fact Cards for 7
- Multiplication Chart
- Multiplication Chart — 6
- Multiplication Chart — 7
- Multiplication Chart — 8
- Multiplication Chart — 9

Activities

Fewer games and activities are included in this chapter as students will be using measuring tools. The included activities can be used after students complete the **Do** questions, or anytime additional practice is needed.

Chapter Opener

Objective

- Investigate multiplication and division of 6, 7, 8, and 9.

Lesson Materials

- Multiplication Chart (BLM)

Have students complete the facts that they already know on the Multiplication Chart (BLM) and discuss patterns on the chart.

Have students think about how to apply strategies they already know to learn the facts that they do not know.

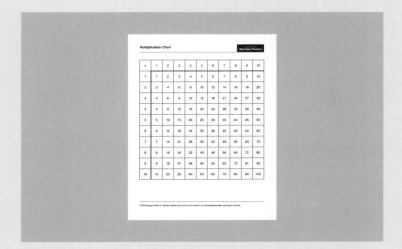

Lesson 1 The Multiplication Table of 6

Objective

- Use the properties of operations to derive new facts from known facts in the multiplication table of 6.

Lesson Materials

- Array Dot Cards — 6 (BLM), 1 set per student
- Multiplication Chart — 6 (BLM)
- Index cards

Think

Pose the **Think** problem and provide each student with a set of Array Dot Cards — 6 (BLM). Ask students how they can build all the facts for 6 with the dot cards.

For example, a student may notice that she knows that 5 × 6 is 30, so another 6 is 30 + 6:

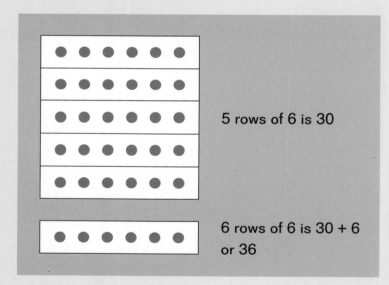

5 rows of 6 is 30

6 rows of 6 is 30 + 6 or 36

Hand out a Multiplication Chart — 6 (BLM) to each student and have them complete the chart for 6.

Discuss Alex's comment and patterns that students notice.

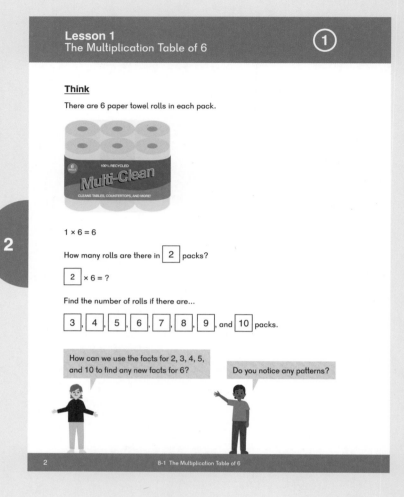

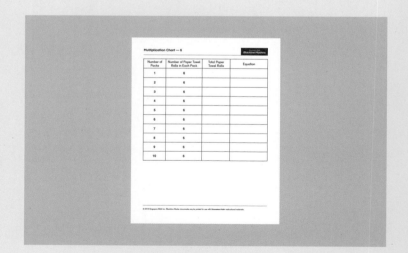

Teacher's Guide 3B Chapter 8

Learn

Have students compare the products and equations on their charts with the chart shown under the **Learn** header. Two strategies for finding unknown facts are shown in this chart.

Discuss any additional strategies students may think of for finding unknown facts from known facts. For example, if students subtract 1 group, the product decreases by 6. As groups increase or decrease by 1, the product increases or decreases by 6. Thus, an easy way to find 9 × 6 is to subtract 6 from 10 × 6.

Have students discuss any patterns they notice in the numbers in the multiplicand or the product. Two patterns are given by Sofia and Mei.

Mei states that the products are even. Remind students that in Dimensions Math 3A Chapter 4: Multiplication and Division, they learned about even numbers, and that the product of odd and even, and even and even, was an even number. This can help students realize that all answers to multiplication × 6 must be an even number; an odd number for the product would be incorrect.

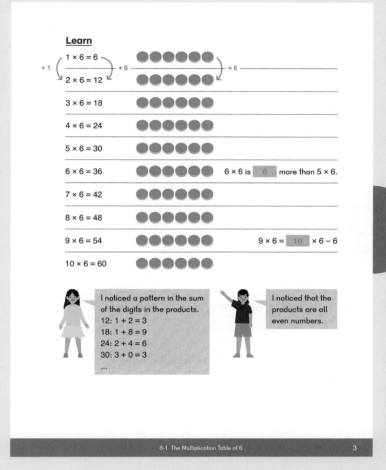

Do

Students should see that they can determine these products from facts they have already learned. By splitting one of the numbers being multiplied into smaller numbers, students can then add the product of facts they know to find the product of a fact they do not know. This is shown in the thought bubbles with number bonds.

This is a strategy students can use before they have memorized the additional facts, or when they have forgotten a fact; it is quicker than successively adding 6 (skip counting by 6).

❶ Emma splits 6 into 5 and 1. She knows 5 × 6 = 30 and adds one more group of 6.

❷ Alex knows that 7 is the sum of 5 and 2. He adds the products of 5 × 6 and 2 × 6 to find the product of 7 × 6.

❸ Mei knows 8 × 6 is double 4 × 6. She adds 2 groups of 4 × 6 or 24 + 24.

As students look at Textbook pages 4 and 5, ask them to focus on the numbers that are bolded. For example in ❷, 5, 2 and 7 are bolded. Add the products of 5 × 6 and 2 × 6 to find the product of 7 × 6.

❺ Discuss Sofia's comment on facts for 6. Have students fold the dot grid on Multiplication Chart — 6 (BLM) to see that facts for 6 are double facts for 3.

Students may also notice that the second number in the problem is bolded this time, and the first number does not change. In this case, the second number in the multiplication fact is split. You can point out to students that they can split it a different way. For example, they could sum the products of 8 × 2 and 8 × 4 to find 8 × 6.

❻ This problem reminds students that the answer is the same regardless of the order of the factors. Point out that they now know 3 new facts, one each for × 7, × 8, and × 9, from the × 6 facts.

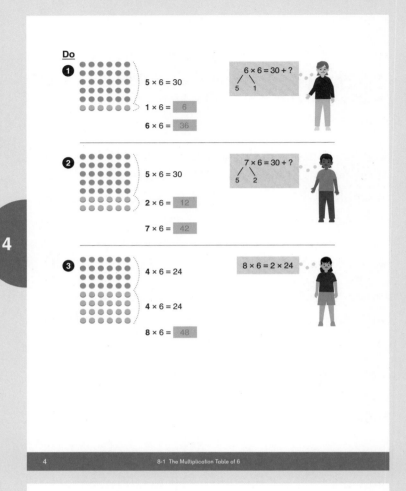

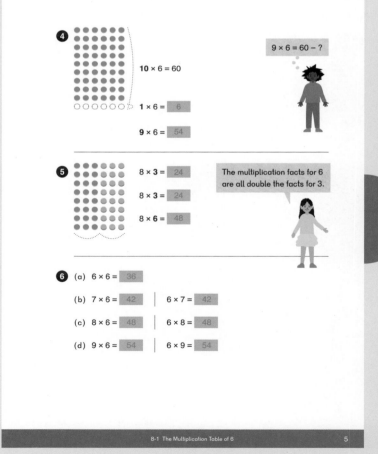

7 This question reminds students pictorially that they can use a related multiplication fact to find the missing number in the division equation.

9 The problems are scaffolded. If students know 24 ÷ 6 = 4, they should know that 28 is 4 more than 24, and therefore 28 ÷ 6 will result in an answer of 4 remainder 4.

Discuss Alex's question about remainders. Students should see that if they have a remainder greater than the number 6, then we can still make another group of 6. In division problems, when we divide a number by 6, the remainder must be less than 6.

Provide students with index cards and have them create their own flash cards for the multiplication table of 6 to use for practice and games.

Activity

▲ Multiplication Wheels

Materials: Paper plates with the center cut out

Create several multiplication wheels with the numbers 1 to 10 in random order as "spokes" along the edge of the paper plates.

Students lay the wheel on a whiteboard and write the number they are multiplying by in the center of the wheel (in this lesson it is 6).

Students multiply the number on the spoke and the number in the center, and write the product on the whiteboard, outside of the wheel.

Exercise 1 • page 1

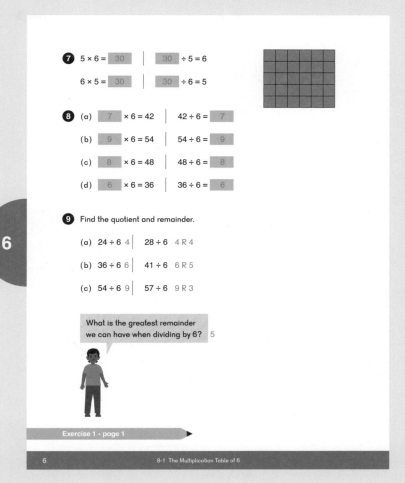

Lesson 2 The Multiplication Table of 7

Objective

- Use the properties of operations to derive new facts from known facts in the multiplication table of 7.

Lesson Materials

- Array Dot Cards — 7 (BLM), 1 set per student
- Multiplication Chart — 7 (BLM)
- Multiplication Chart (BLM)
- Index cards

Think

Pose the **Think** problem and provide each student with a set of Array Dot Cards — 7 (BLM).

Ask students:

- How is this similar to the **Think** problem in the prior lesson?
- How can we find the facts for the multiplication table of 7 that we do not know?

Provide each student with a Multiplication Chart — 7 (BLM) and have them record the multiplication table for 7.

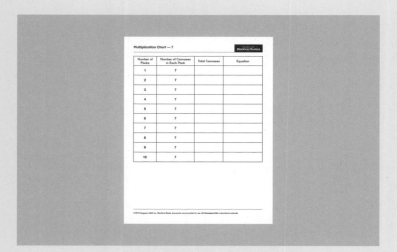

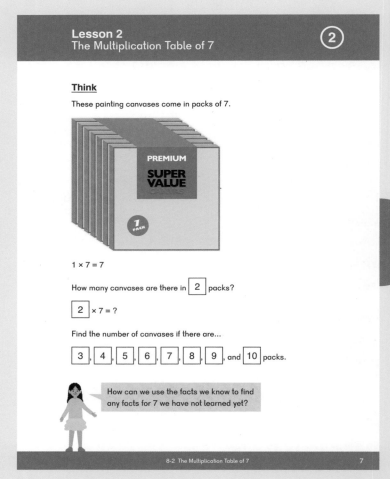

10   Teacher's Guide 3B Chapter 8   © 2017 Singapore Math Inc.

Learn

Ask students:

- How was this problem different from the problem we did yesterday? (We are multiplying by 7.)
- How was it the same? (We are still making equal groups.)

Have students compare the products and equations on their charts with the one shown in the textbook.

Have students discuss any patterns they notice in the numbers in the multiplicand or the product.

If students know 6 × 7, they can find 7 × 7 by simply adding 7.

Similarly, if the first number decreases by 1, the product decreases by 7. They can use this idea to find 9 × 7 from 10 × 7.

Mei points out that students really have only 3 new facts to learn.

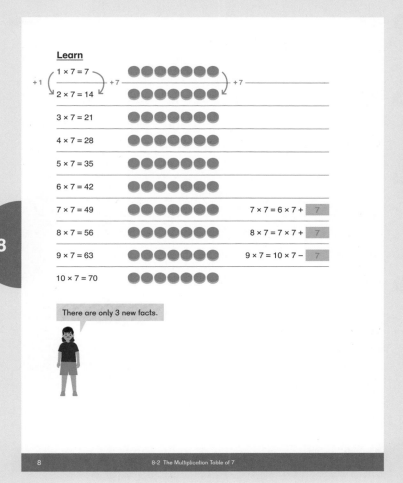

Do

1. If students know 3 × 7 = 21, they can find 6 × 7 by adding together 2 groups of 3 × 7 or 21 + 21.

   Students can confirm that this is true by using a Multiplication Chart (BLM) and comparing the products in the row for × 3 and × 6. (While doing this, students may note other patterns not taught, such as the × 8 products are double × 4 products and the × 10 products are double × 5 products.)

2. Students should see that if they know the products of 5 × 7 and 2 × 7, they can find the product of 7 × 7 and determine that 7 sevens is 2 more sevens than 5 sevens.

3. Emma splits the multiplier (7) instead of the multiplicand. Students should notice that one of the factors stays the same, this time the one on the left (8), while the one on the right (7) is split.

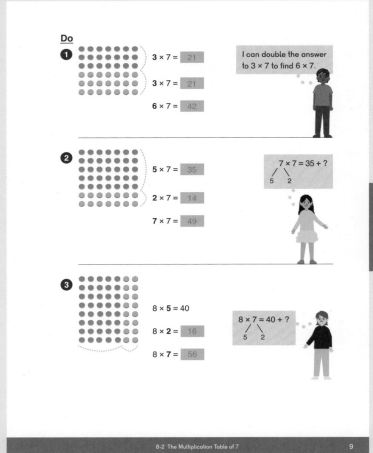

6 – 7 These problems remind students that they can think of a related multiplication fact to find the answer to the facts for division, with **6** showing this pictorially.

8 – 9 Ask students to write the equations and explain how they knew whether to multiply or divide. Students can draw bar models as needed.

8

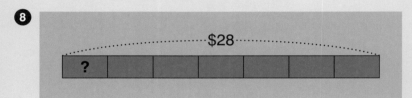

9

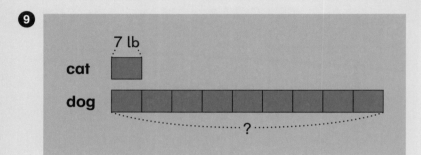

10 Students can use the problem on the left to help them determine what the remainder is for the problem on the right. Reinforcing what they learned in the previous lesson about dividing by 6, Mei reminds them that the remainder has to be less than the number they are dividing by.

Provide students with index cards and have them create their own flash cards for the multiplication table of 7. Have them use these flash cards for practice and games.

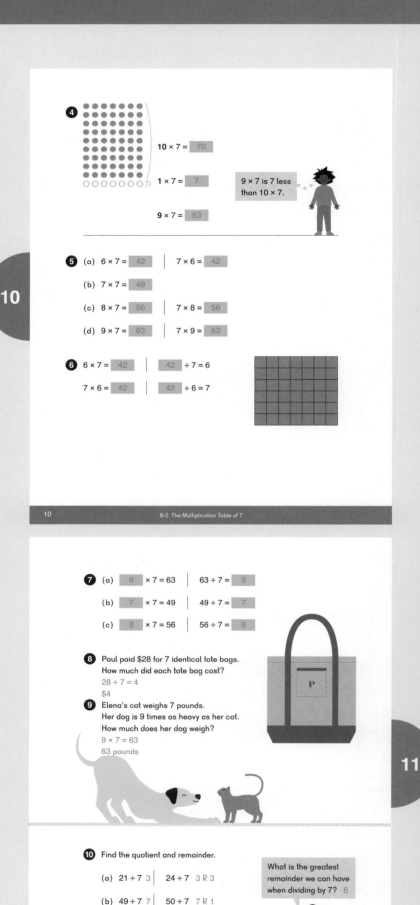

12   Teacher's Guide 3B Chapter 8   © 2017 Singapore Math Inc.

Activities

▲ Multiplication Wheels

Materials: Paper plates with the center cut out

Multiplication Wheels from the previous lesson can be created and used for any facts students need to practice.

▲ Fact or Not a Fact?

Give students a random number and ask if it is a product of 6 or 7.

For example:

- Is 41 a product of 7?
- Is 37 a product of 6?

▲ Choral Counting

Pointing your thumb up or down, have students chorally count up and down by sevens.

Example: "Let's count by sevens starting at 0. First number?" The class responds, "0." Then, point your thumb up, and the class responds, "7." Then point your thumb up again, and the class responds, "14." Point down, and the class responds, "7," and so on.

Start at random multiples of 7. "Let's count by sevens starting at 28. First number?" The class responds, "28." Then, point your thumb up, and the class responds, "35." Points your thumb up again, and the class responds, "42." Point down, and the class responds, "35," and so on.

Exercise 2 • page 4

Lesson 3 Multiplying by 6 and 7

Objective
- Multiply a number of up to 3 digits by 6 or 7.

Lesson Materials
- Place-value discs (if needed)

Think

Pose the **Think** problem and have students try to solve the problem independently.

Have students share and discuss how they solved the problem.

Learn

Discuss the steps in the multiplication algorithm with students. Note that place-value discs are not used in the textbook. If students struggle, demonstrate the algorithm for them with place-value discs.

For students who still struggle, the partial products strategy may provide a bridge to the algorithm.

```
    2 7 5
  ×     6
       30  ← 5 × 6
      420  ← 70 × 6
    1,200  ← 200 × 6
    1,650
```

Have students compare their solutions from **Think** with the one shown in the textbook.

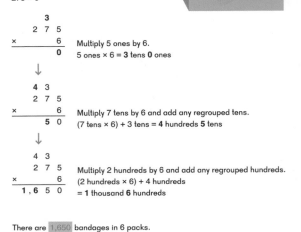

14 Teacher's Guide 3B Chapter 8 © 2017 Singapore Math Inc.

Do

1 — **2** Ensure students understand the steps in the multiplication algorithm for each problem.

1 Ask students:

- What is 6 × 9 ones? (54)
- Where do we write the 4 ones and 5 tens?
- What do we multiply by next? (6 tens by 6.)
- What is the next step? (Add the 5 tens.)
- Where do we write the answer?

5 While the question asks to find the product of 258 groups at $7 a group, it may be easier to think of the calculation as 7 groups of 258. Since multiplication is commutative, the result is the same.

Students should be comfortable multiplying using either multiple groups or multiple quantities per group.

Exercise 3 • page 7

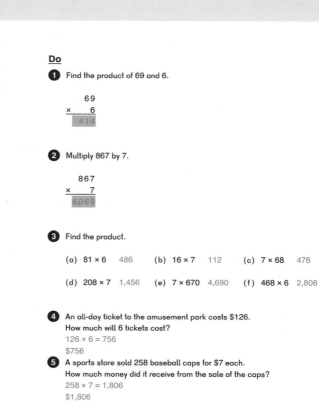

Lesson 4 Dividing by 6 and 7

Objective

- Divide a number of up to 3 digits by 6 and 7.

Think

Pose the **Think** problem and have students try to solve the problem independently.

Have students share and discuss how they solved the problem.

Some students may notice that instead of dividing the eggs into 6 groups, we are dividing into groups of 6. Students who struggle to solve the problem because of this should be reminded that the procedure works whether we divide into groups of 6 or 6 groups.

Learn

Discuss the steps in the division algorithm with students.

Have students compare their solutions from **Think** with the one shown in the textbook.

Do

① Ensure students understand the steps in the division algorithm for each problem.

(a) Dion points out that the final remainder should be less than 6. As students do the problem, the digit for each step in the algorithm showing the remainder for each place should be less than 6. Otherwise, the answer when dividing the next lower place will not be a single digit.

Mei reminds students that when checking their work they can multiply the quotient by the divisor, and add the remainder to get the dividend (number being divided).

③ — ④ Students can draw bar models to help them solve these problems. Examples:

③
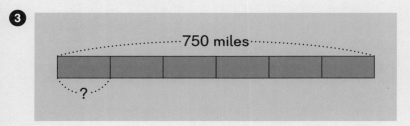

④

Exercise 4 • page 10

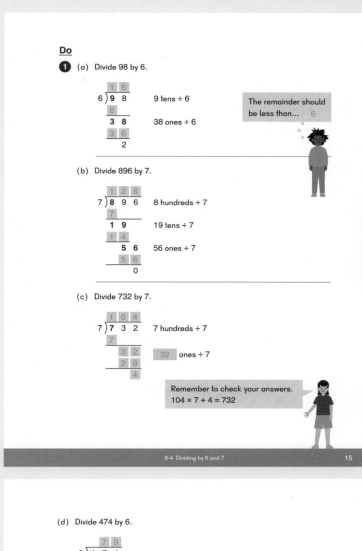

Lesson 5 Practice A

Objectives

- Practice multiplying and dividing by 6 and 7.
- Solve word problems involving multiplication and division.

After students complete the **Practice** in the textbook, have them continue to practice multiplying and dividing by 6 and 7 using activities from this chapter.

3 Students could also draw a bar model.

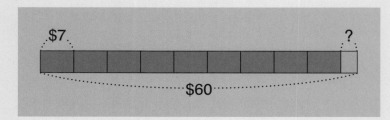

5 Students can also solve the problem using equations, with a letter standing for the unknown quantity:

w = number of weeks
x = number of weeks that remain

w = 259 ÷ 7
x = 52 − w

Activity

▲ **Multiplication and Division Kaboom**

Materials: 3 Kaboom Cards (BLM), several sets of Multiplication and Division Fact Cards for 6 (BLM), Multiplication and Division Fact Cards for 7 (BLM)

Shuffle and place the cards in a pile facedown. Players take turns drawing a card and answering the multiplication or division fact.

Students keep the cards they answer correctly, and return the ones that they answer incorrectly. When a student draws a Kaboom Card (BLM), he must return all of his collected cards to the pile.

The player with the most cards at the end of the time limit is the winner.

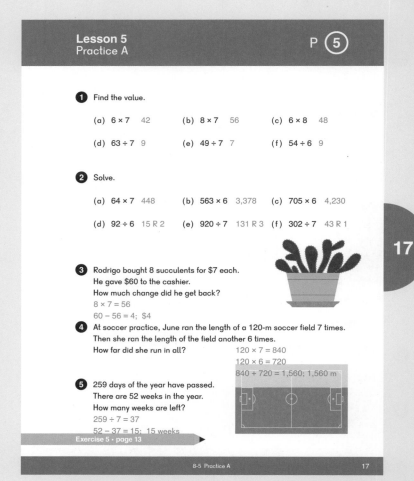

Lesson 6 The Multiplication Table of 8

Objective

- Use the properties of operations to derive new facts from known facts in the multiplication table of 8.

Lesson Materials

- Multiplication Chart — 8 (BLM)
- Index cards

Think

Pose the **Think** problem and provide each student with a Multiplication Chart — 8 (BLM).

Ask students:

- How is this similar to the multiplication tables of 6 and 7?
- How is this similar to the multiplication tables of 2 and 4?
- How can we find the facts for the multiplication table of 8 that we do not know?

Have students complete the multiplication table on Multiplication Chart — 8 (BLM).

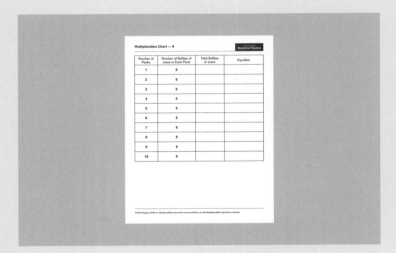

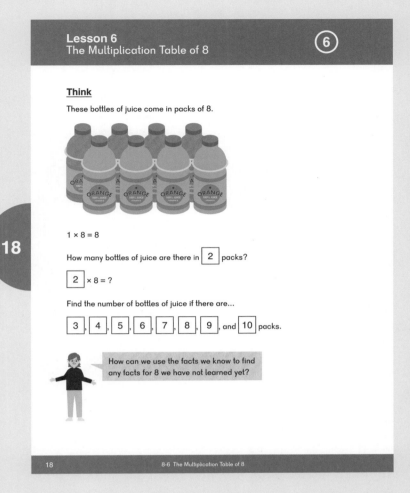

Learn

Have students compare the products and equations on their charts with the one shown in the textbook.

The strategies shown here are the same as previously shown for other multiplication tables.

Sofia points out that there are only two new facts to learn: 8 × 8 and 8 × 9.

Discuss Alex's comment. Students can see that going down the table, the digits in the ones place decrease by 2. This can be understood by thinking about the "over-adding" strategy of adding 8 by adding 10 and subtracting 2.

Do

② Dion knows that 7 is the sum of 5 and 2. He adds the products of 5 × 8 and 2 × 8 to find the product of 7 × 8.

④ To find 8 × a number, first find twice the number and then double that product. (Students may note that 2 × 2 × 2 = 8.)

For example, to find 8 × 5, first find 2 × 5, then double the product (10) to get 20, and double 20 to get 40.

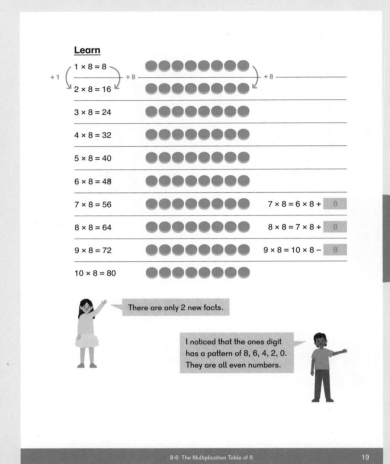

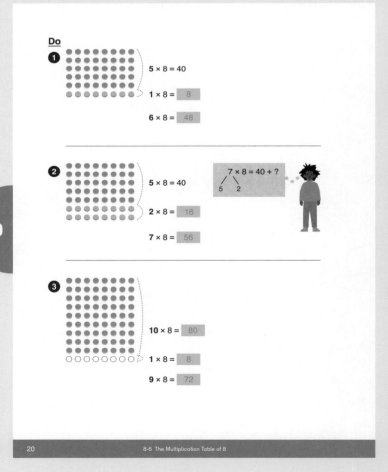

8 (a) Students could think:

- What number + 2 = 74?
- What number × 8 gives a product of 72?

(b) Students could think:

- What number + 6 = 54?
- What number × 8 gives a product of 48?

Provide students with index cards and have them create their own flash cards for the multiplication table of 8 to use for practice and games.

Activities

Three of the activities from Lessons 1 and 2 can be used in this lesson:

▲ **Multiplication Wheels**

▲ **Choral Counting**

▲ **Fact or Not a Fact?**

▲ **Multiplication Team Race**

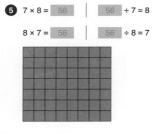

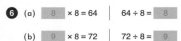

Students seeking a challenge will enjoy the race element of this game. Have 2–4 students go to the board and make a large "T" shape. Randomly call out 5 numbers from 1 to 10 and have students write them on the left side of the T.

Then give them a number to multiply by, for example, "times 8" in this scenario. Students write "× 8" on the top of the T, then proceed to solve the problems on the right side as fast as they can.

Exercise 6 • page 17

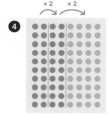

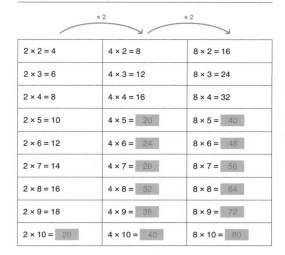

Lesson 7 The Multiplication Table of 9

Objective

- Use the properties of operations to derive new facts from known facts in the multiplication table of 9.

Lesson Materials

- Multiplication Chart — 9 (BLM)
- Index cards

Think

Pose the **Think** problem and provide each student with a Multiplication Chart — 9 (BLM).

Ask students:

- How is this similar to other multiplication tables?
- How can we find the facts for the multiplication table of 9 that we do not know?

Have them complete the multiplication table on Multiplication Chart — 9 (BLM).

- 1 package, 9 light bulbs in each package, total number of light bulbs

Discuss Mei and Alex's comments. The only new fact is 9 × 9.

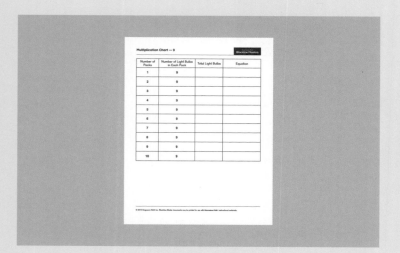

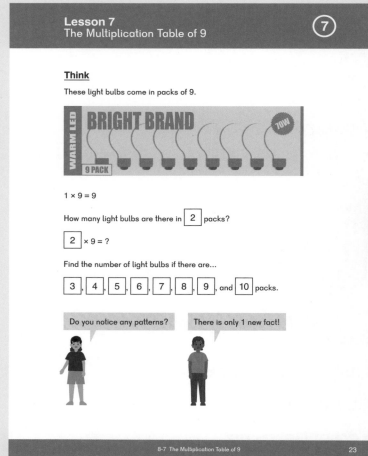

Learn

Have students compare the products and equations on their charts with the one shown in the textbook.

The strategies shown here are the same as previously shown for other multiplication tables.

Emma and Dion's thoughts go together to provide an idea of an easy way to find the product. First, the tens digit is one less. Next, the ones digit is what you would add to the tens digit to get 9.

Emma notices that the tens digit is one less than the number of groups.

Discuss Dion's thought. He notices that the sum of the digits in the product is 9:

$6 \times 9 = 54$ and $5 + 4 = 9$.

Do

1. Discuss the different ways the facts for 9 are found.

2. Adding 9 is 1 less than adding 10. Thus, if students know $5 \times 9 = 45$, to find 6×9, they can add 10 to 45 and subtract 1.

$6 \times 9 = 5 \times 9 + 10 - 1$

Mei notices that as the tens digit in the product increases by 1, the ones digit decreases by 1.

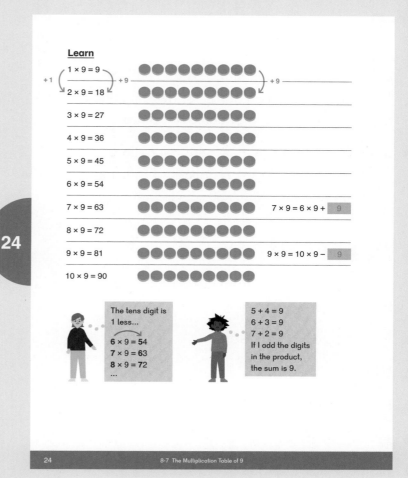

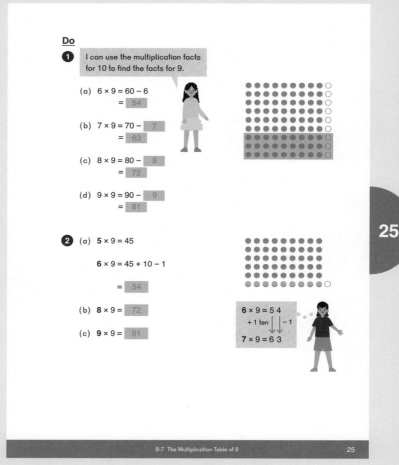

5 — 6 Ask students to write the equations and explain how they knew whether to multiply or divide.

Provide students with index cards and have them create their own flash cards for the multiplication table of 9 to use for practice and games.

5 Students should see that there needs to be an additional driver for the remainder 6 passengers, so 7 drivers are needed for 60 passengers.

Activities

Four of the activities from Lessons 1, 2, and 6 can be used in this lesson:

▲ **Multiplication Wheels**

▲ **Choral Counting**

▲ **Fact or Not a Fact?**

▲ **Multiplication Team Race**

▲ **Rock, Paper, Scissors, Math!**

Players bounce a fist in one hand while saying, "rock, paper, scissors, math."

On the word "math," each player shoots out some fingers on both hands. The player who says the product of the fingers first is the winner.

For example, if Player One shows 7 fingers, and Player Two shows 6 fingers, the first player to say, "42," wins.

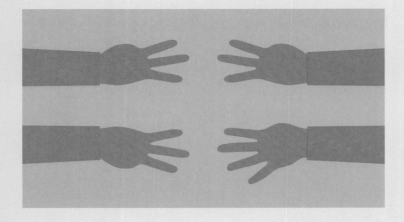

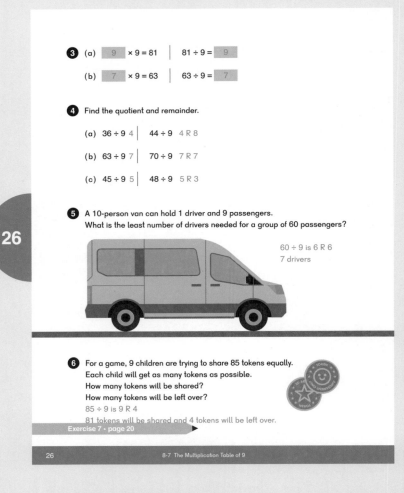

3 (a) 9 × 9 = 81 | 81 ÷ 9 = 9
  (b) 7 × 9 = 63 | 63 ÷ 9 = 7

4 Find the quotient and remainder.
  (a) 36 ÷ 9 4 | 44 ÷ 9 4 R 8
  (b) 63 ÷ 9 7 | 70 ÷ 9 7 R 7
  (c) 45 ÷ 9 5 | 48 ÷ 9 5 R 3

5 A 10-person van can hold 1 driver and 9 passengers. What is the least number of drivers needed for a group of 60 passengers?
60 ÷ 9 is 6 R 6
7 drivers

6 For a game, 9 children are trying to share 85 tokens equally. Each child will get as many tokens as possible. How many tokens will be shared? How many tokens will be left over?
85 ÷ 9 is 9 R 4
81 tokens will be shared and 4 tokens will be left over.

Exercise 7 • page 20

Lesson 8 Multiplying by 8 and 9

Objective

- Multiply a number of up to 3 digits by 8 or 9.

Think

Pose the **Think** problem and have students try to solve the problem independently.

Discuss students' strategies for solving the problem.

Learn

Discuss the steps in the multiplication algorithm with students.

Emphasize the place value of the number multiplied: we multiply 7 tens by 8 and have 56 tens, which we regroup.

Have students compare their solutions from **Think** with the one shown in the textbook.

Do

4 – **5** Ask students to write the equations and explain how they knew whether to multiply or divide.

5 From the bar model, students can see two different ways of solving the problem.

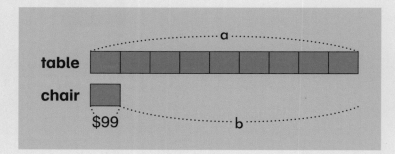

Students can also solve the problem using equations, with a letter standing for the unknown quantity:

a = the cost of the table
b = the difference in cost

a = $99 × 9
b = a − $99

Exercise 8 • page 23

Lesson 9 Dividing by 8 and 9

Objective

- Divide a number of up to 3 digits by 8 and 9.

Think

Pose the **Think** problem and have students try to solve the problem independently.

Discuss student strategies for solving the problem.

Some students may notice that instead of dividing the donuts into 9 groups, we are dividing into groups of 9. Students who struggle to solve the problem because of this should see that the procedure works whether we divide into groups of 9 or 9 groups.

Learn

Discuss the steps in the division algorithm with students.

Have students estimate their answers. 954 is more than 900. I can estimate 900 ÷ 9 is 100, so my quotient for 954 ÷ 9 will be greater than 100.

As 954 splits nicely into multiples of 9, students may also use a mental strategy:

$$954 \div 9 = 100 + 6$$
$$\diagup \quad \diagdown$$
$$900 \quad 54$$

Remind students that they can check their division by multiplying the quotient and divisor.

Have students compare their solutions from **Think** with the one shown in the textbook.

Do

1 Ensure students understand the steps in the division algorithm for each problem.

3 Note that the question is asking for the number of children not on a team, which would be the remainder.

Exercise 9 • page 26

Lesson 10 Practice B

Objectives

- Practice multiplying and dividing by 8 and 9.
- Solve word problems involving multiplication and division.

After students complete the **Practice** in the textbook, have them continue practicing multiplication and division facts for 6, 7, 8, and 9. Students should continue practicing these facts until they know them from memory.

Activity

▲ **Snowball Review**

Materials: Blank paper

Write multiplication and division problems on sheets (or half sheets) of paper. Make enough to give each student 2 or 3 problems.

Have students crumple up their papers and have a classroom snowball fight for one minute. (No running, safety first!)

At the end of the fight, ask kids to grab a snowball and return to their seats. One at a time, have students unwrap the paper, show the class, and state the answer to the problem.

After each student has solved one fact problem, have another snowball fight and repeat with remaining snowballs on the floor.

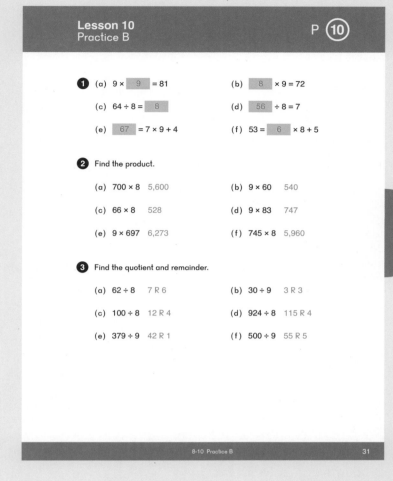

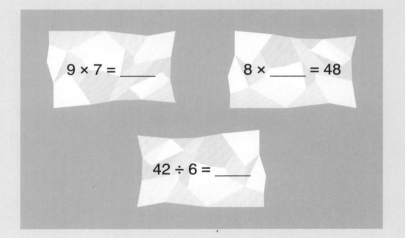

④ From the given model, students can see that Taylor saved $18 ÷ 2, or $9, and that each unit is therefore equal to $9.

Students can then solve for the difference in the two girls' savings in two ways:

(a) Find Yara's savings ($9 × 9), and then subtract $9 (Taylor's savings) to find the difference.

(b) Find the difference in the number of units (9 − 1 = 8), and multiply by the value of each unit:

1 unit ⟶ $9
8 units ⟶ $9 × 8 = $72

⑤
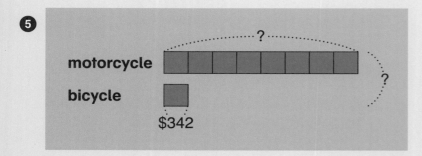

Students can solve in two ways:

(a) Find the cost of the motorcycle first, then add that to the cost of the bicycle.
(b) Multiply the cost of the bicycle by 9:

1 unit ⟶ $342
9 units ⟶ $342 × 9 = $3,078

⑥
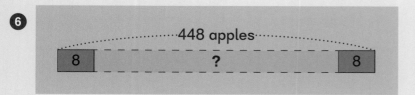

First, find the number of bags of apples:

448 ÷ 8 = 56 bags

Next, find how much money the store received:

56 × $9 = $504

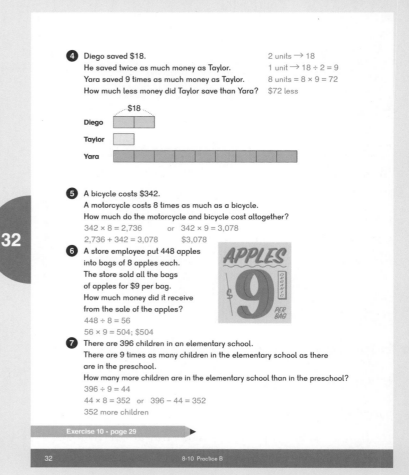

⑦
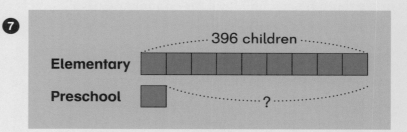

First, divide 396 by 9 to find the number of preschool students, which represents the value of a single unit.

9 units ⟶ 396 children
1 unit ⟶ 396 ÷ 9 = 44 children

Next, students can arrive at the final answer by using one of two methods:

(a) Subtract: 396 − 44 = 352 children
(b) Multiply: 8 units × 44 = 352 children

Exercise 10 • page 29

Exercise 1 • pages 1–3

Chapter 8 Multiplying and Dividing with 6, 7, 8, and 9

Exercise 1

Basics

1. (a) 6 × 6 is 6 more than [5] × 6.

    6 × 6 = [36]

    (b) 9 × 6 is 6 less than [10] × 6.

    9 × 6 = [54]

2. (a) 8 × 6 = [24] + 24 = [48]

    (with 4, 4 below)

    (b) 7 × 6 = 30 + [12] = [42]

    (with 5, 2 below)

3. 6 × 9 = 9 × [6] = [54]

4. (a) [9] × 6 = 54 | 54 ÷ 6 = [9]

    (b) [6] × 6 = 36 | 36 ÷ 6 = [6]

5. 42 ÷ 6 = [7] | 42 + [3] = 45

    45 ÷ 6 is [7] with a remainder of [3].

Practice

6. Find the missing numbers.

4 × 6 = [24] **P**	18 ÷ 6 = [3] **C**	10 × 6 = [60] **T**
54 ÷ 6 = [9] **X**	0 × 6 = [0] **E**	7 × 6 = [42] **O**
6 × 2 = [12] **I**	42 ÷ 6 = [7] **G**	5 × 6 = [30] **H**
6 × 8 = [48] **D**	6 × [2] = 12 **S**	48 ÷ 6 = [8] **K**
6 × 3 = [18] **E**	36 ÷ 6 = 6 **C**	9 × 6 = [54] **N**
6 × [5] = 30 **S**	1 × [6] ÷ 6 **A**	[4] = 24 ÷ 6 **.**

What are three animals to first be domesticated for food?
Write the letters that match the answers above to find out.

G	O	A	T	S	,	S	H	E	E	P	,
7	42	6	60	5	4	15	2	30	0	18	24	4

A	N	D		C	H	I	C	K	E	N	S
6	54	48	20	3	30	12	36	8	18	54	5	56

7. (a) 8 × 6 = 6 + 6 + 6 + [30]

    (b) 9 × 4 = 6 × [6]

8. (a) 23 ÷ 6 is [3] with a remainder of [5].

    (b) 59 ÷ 6 is [9] with a remainder of [5].

    (c) 50 ÷ 6 is [8] with a remainder of [2].

9. Connor bought 6 packs of party favors.
   Each pack came with 8 party favors.
   He gave out 37 party favors.
   How many party favors does he have left over?

   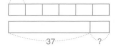

   6 × 8 = 48 (total party favors)
   48 − 37 = 11 (left over)
   He has 11 party favors left over.

Challenge

10. Aiden has $75.
    He spent $28 on a pair of pants and $6 each on some shirts.
    He has $5 left over.
    How many shirts did he buy?

    | $5 | $28 | $6 | ? | $6 |

    $75

    75 − 5 = 70 (amount spent)
    70 − 28 = 42 (amount spent on shirts)
    42 ÷ 6 = 7
    He bought 7 shirts.

Exercise 2 • pages 4–6

Exercise 2

Basics

1. 9 × 7 is [7] less than 10 × 7.
   9 × 7 = [63]

2. (a) 7 × 7 = [35] + 14 = [49]
       (5, 2)
   (b) 6 × 7 = [21] + 21 = [42]
       (3, 3)
   (c) 8 × 7 = 40 + [16] = [56]
       (5, 2)

3. 7 × 8 = 8 × [7] = [56]

4. (a) [9] × 7 = 63   63 ÷ 7 = [9]
   (b) [4] × 7 = 28   28 ÷ 7 = [4]

5. 56 ÷ 7 = [8]   56 + [6] = 62
   62 ÷ 7 is [8] with a remainder of [6].

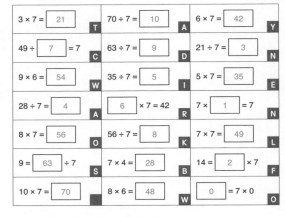

Practice

6. Find the missing numbers.

3 × 7 = [21] T	70 ÷ 7 = [10] A	6 × 7 = [42] Y
49 ÷ [7] = 7 C	63 ÷ 7 = [9] D	21 ÷ 7 = [3] N
9 × 6 = [54] W	35 ÷ 7 = [5] I	5 × 7 = [35] E
28 ÷ 7 = [4] A	[6] × 7 = 42 R	7 × [1] = 7 N
8 × 7 = [56] O	56 ÷ 7 = [8] K	7 × 7 = [49] L
9 = [63] ÷ 7 S	7 × 4 = [28] B	14 = [2] × 7 F
10 × 7 = [70]	8 × 6 = [48]	[0] = 7 × 0 O

Joke: Why was the clock in the cafeteria always slow?
Write the letters that match the answers above to find out.

	I	T		A	L	W	A	Y	S
17	5	21	70	4	49	54	10	42	63	33

	W	E	N	T		B	A	C	K
44	48	35	1	21	70	28	4	7	8	19

F	O	R		S	E	C	O	N	D	S
2	56	6	70	63	35	7	0	3	9	63

7. (a) 8 × 7 = 7 + 7 + [42]
   (b) 9 × 7 = 7 × [4] + 35

8. (a) 65 ÷ 7 is [9] with a remainder of [2].
   (b) 50 ÷ 7 is [7] with a remainder of [1].
   (c) 62 ÷ 7 is [8] with a remainder of [6].
   (d) 30 ÷ 7 is [4] with a remainder of [2].
   (e) 48 ÷ 7 is [6] with a remainder of [6].

9. There are 35 days in [5] weeks.
   35 ÷ 7 = 5

10. There are [63] days in 9 weeks.
    9 × 7 = 63

Challenge

11. A box of 10 gel pens costs $7.
    Grace spent $56 on boxes of these pens.
    Then she gave 7 pens to each of her friends.
    She has 38 pens left.
    How many friends did she give pens to?
    56 ÷ 7 = 8 (number of boxes of pens)
    8 × 10 = 80 (total number of pens)
    80 − 38 = 42 (number of pens she gave away)
    42 ÷ 7 = 6
    She gave pens to 6 friends.

Exercise 3 • pages 7–9

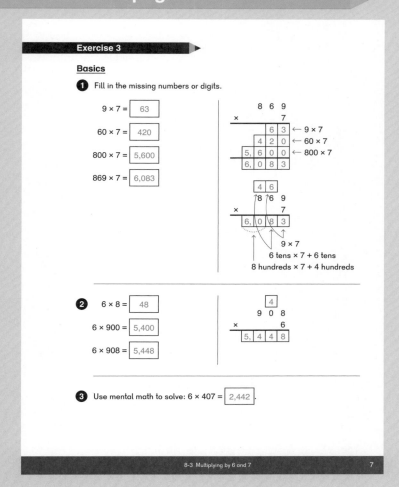

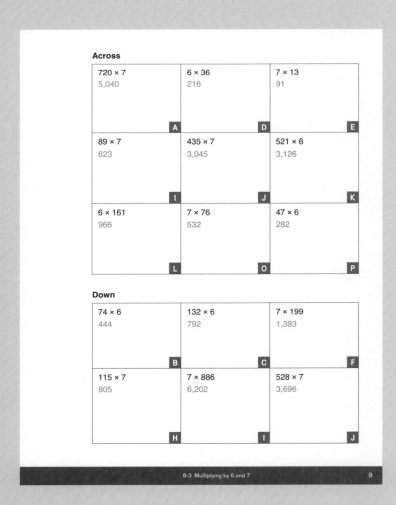

Exercise 4 • pages 10–12

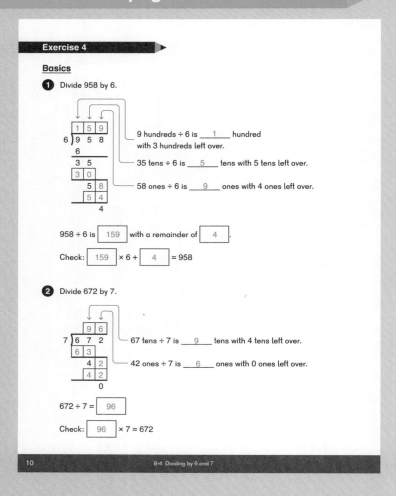

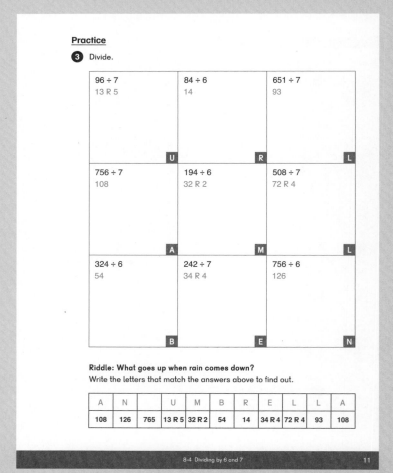

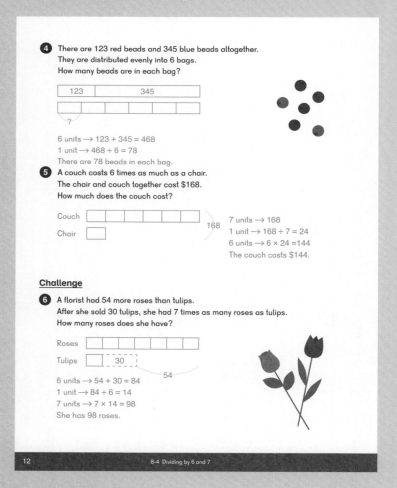

Teacher's Guide 3B Chapter 8 © 2017 Singapore Math Inc.

Exercise 5 • pages 13–16

Exercise 5

Check

1 The product of the two numbers in the overlapping squares is the third number in each large square.
Fill in the missing numbers.

(a)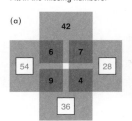

(b)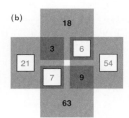

(c)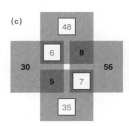

(d)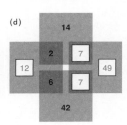

2 Multiply or divide.

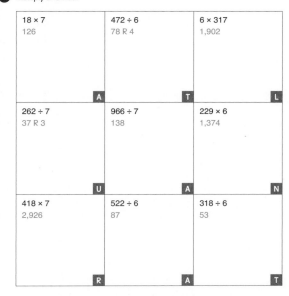

What spider from the Amazon Basin is the largest type of spider in the world?
Write the letters that match the answers above to find out.

A	T	A	R	A	N	T	U	L	A
126	527	78 R 4	87	2,926	138	1,374	53	37 R 3	1,902	126

3 Write the missing digits.

(a)

(b)

(c)

(d)

4 2 identical bikes cost $396.
How much do 7 such bikes cost?

2 units ⟶ 396
1 unit ⟶ 396 ÷ 2 = 198
7 units ⟶ 198 × 7 = 1,386
7 bikes cost $1,386.

5 Emiliano wants to pack 320 candles equally into 6 identical boxes with the fewest number of candles left over.
How many candles will be left over?
320 ÷ 6 is 53 R 2
There will be 2 candles left over.

6 Brianna has twice as much money as Amanda.
Catalina has twice as much money as Brianna.
Altogether, they have $595.
How much money does Catalina have?

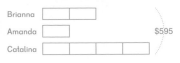

7 units ⟶ 595
1 unit ⟶ 595 ÷ 7 = 85
4 units ⟶ 4 × 85 = 340
Catalina has $340.

Challenge

7 The product of the two numbers in the overlapping squares is the third number in each large square.
Fill in the missing numbers.

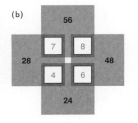

Teacher's Guide 3B Chapter 8

Exercise 6 • pages 17–19

Exercise 6

Basics

1. $6 \times 8 = 8 \times 6 = $ 48

2. $9 \times 8 = 10 \times 8 - $ 8 $ = $ 72

3. $7 \times 8 = $ 40 $ + 16 = $ 56
   (5, 2)

4. (a) $8 \xrightarrow{\times 2} 16 \xrightarrow{\times 2} 32 \xrightarrow{\times 2} 64$
   $8 \times 8 = 64$

   (b) $4 \xrightarrow{\times 2} 8 \xrightarrow{\times 2} 16 \xrightarrow{\times 2} 32$
   $4 \times 8 = 32$

5. (a) 9 $\times 8 = 72$ | $72 \div 8 = $ 9
   (b) 3 $\times 8 = 24$ | $24 \div 8 = $ 3

6. $64 \div 8 = $ 8 | $64 + $ 6 $ = 70$
   $70 \div 8$ is 8 with a remainder of 6.

Practice

7. Multiply or divide.

$10 \times 8 = $ 80	$2 \times 8 = $ 16	$72 \div 8 = $ 9
$16 \div 8 = $ 2	$64 \div 8 = $ 8	$40 \div 8 = $ 5
$3 \times 8 = $ 24	$32 \div 8 = $ 4	$80 \div 8 = $ 10
$56 \div 8 = $ 7	$7 \times 8 = $ 56	$0 \times 8 = $ 0
$4 \times 8 = $ 32	$8 \div 8 = $ 1	$6 \times 8 = $ 48
$5 \times 8 = $ 40	$24 \div 8 = $ 3	$8 \times 8 = $ 64
$20 \times 8 = $ 160	$48 \div 8 = $ 6	$9 \times 8 = $ 72

How many teeth do adult elephants have?
Color the spaces that contain the answers to find out.

38	3	80	18	0	17	65
24	17	6	44	4	95	29
11	14	10	34	16	12	35
28	32	150	84	72	5	64
8	51	22	33	160	99	7
40	1	2	580	9	56	48
19	38	13	95	27	50	85

8. (a) $65 \div 8$ is 8 with a remainder of 1.
   (b) $50 \div 8$ is 6 with a remainder of 2.
   (c) $60 \div 8$ is 7 with a remainder of 4.
   (d) $22 \div 8$ is 2 with a remainder of 6.

9. A store owner has 75 toys to display.
   He displays 8 toys on each shelf, and the remaining toys on one more shelf.
   How many shelves does he use?
   $75 \div 8$ is 9 R 3
   He uses 10 shelves.

Challenge

10. Caden has 3 times as many toy dinosaurs as Leo.
    Ella has 5 fewer toy dinosaurs than Caden.
    Altogether, they have 44 toy dinosaurs.
    How many toy dinosaurs does Ella have?

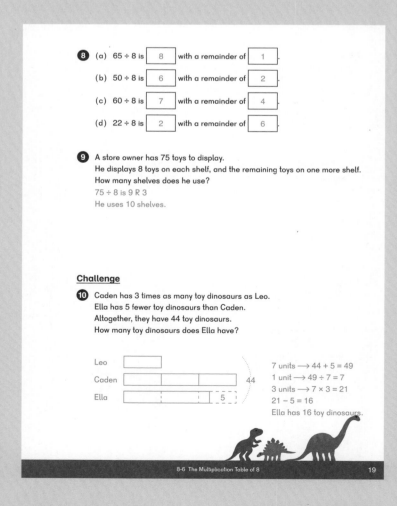

7 units ⟶ 44 + 5 = 49
1 unit ⟶ 49 ÷ 7 = 7
3 units ⟶ 7 × 3 = 21
21 − 5 = 16
Ella has 16 toy dinosaurs.

Exercise 7 • pages 20–22

Exercise 7

Basics

1. (a) $6 \times 9 = 60 - 6 = \boxed{54}$

   (b) $7 \times 9 = 70 - \boxed{7} = 63$

   (c) $8 \times 9 = \boxed{80} - 8 = 72$

   (d) $9 \times 9 = 90 - \boxed{9} = \boxed{81}$

2. (a) $\boxed{5} \times 9 = 45$

   (b) $6 \times 9 = 45 + 10 - \boxed{1} = \boxed{54}$

3. (a) If you add the value of the digits in the products for the multiplication table of 9, when multiplied by numbers up to 10, the sum is $\underline{9}$.

   (b) Circle the numbers that cannot be products of 9 and a whole number.

   (37)   27   36   54   (62)   56

4. (a) $\boxed{9} \times 9 = 81$ | $81 \div 9 = \boxed{9}$

   (b) $\boxed{7} \times 9 = 63$ | $63 \div 9 = \boxed{7}$

5. $63 \div 9 = \boxed{7}$ | $63 + \boxed{6} = 69$

   $69 \div 9$ is $\boxed{7}$ with a remainder of $\boxed{6}$.

Practice

6. Find the missing numbers.

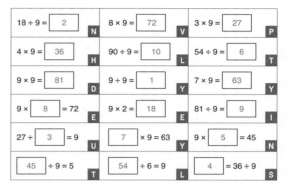

$18 \div 9 = \boxed{2}$ N	$8 \times 9 = \boxed{72}$ V	$3 \times 9 = \boxed{27}$ P
$4 \times 9 = \boxed{36}$ H	$90 \div 9 = \boxed{10}$ L	$54 \div 9 = \boxed{6}$ T
$9 \times 9 = \boxed{81}$ D	$9 \div 9 = \boxed{1}$ Y	$7 \times 9 = \boxed{63}$ Y
$9 \times \boxed{8} = 72$ E	$9 \times 2 = \boxed{18}$ E	$81 \div 9 = \boxed{9}$ I
$27 \div \boxed{3} = 9$ U	$\boxed{7} \times 9 = 63$ Y	$9 \times \boxed{5} = 45$ N
$\boxed{45} \div 9 = 5$ T	$\boxed{54} \div 6 = 9$ L	$\boxed{4} = 36 \div 9$ S

When scientists tried to find a replacement for rubber, what happened? Write the letters that match the answers above to find out.

T	H	E	Y		I	N	V	E	N	T	E	D
6	36	18	1	12	9	2	72	18	5	6	8	81

S	I	L	L	Y		P	U	T	T	Y
56	4	9	10	54	63	82	27	3	45	6	7	90

7. (a) $100 \div 9$ is $\boxed{11}$ with a remainder of $\boxed{1}$.

   (b) $55 \div 9$ is $\boxed{6}$ with a remainder of $\boxed{1}$.

   (c) $89 \div 9$ is $\boxed{9}$ with a remainder of $\boxed{8}$.

   (d) $35 \div 9$ is $\boxed{3}$ with a remainder of $\boxed{8}$.

8. A deck of playing cards has 52 cards. If as many cards as possible are dealt out equally to 9 players, how many cards does each player get?

   $52 \div 9$ is 5 R 7

   Each player gets 5 cards.

Challenge

9. Tomas has $82 more than Franco. If Tomas gives $5 to Franco, he will have 9 times as much money as Franco. How much money do they have altogether?

   Tomas | | | | | | | | 5 |
   Franco | 5 |
              82

   8 units ⟶ 82 − 5 − 5 = 72
   1 unit ⟶ 72 ÷ 8 = 9
   10 units ⟶ 10 × 9 = 90
   They have $90 altogether.

Exercise 8 • pages 23–25

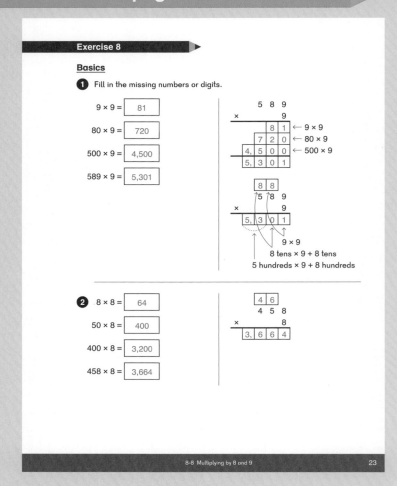

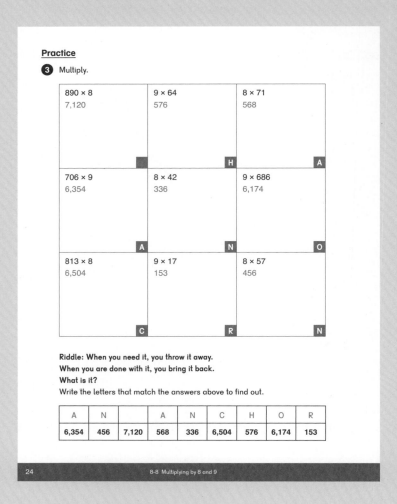

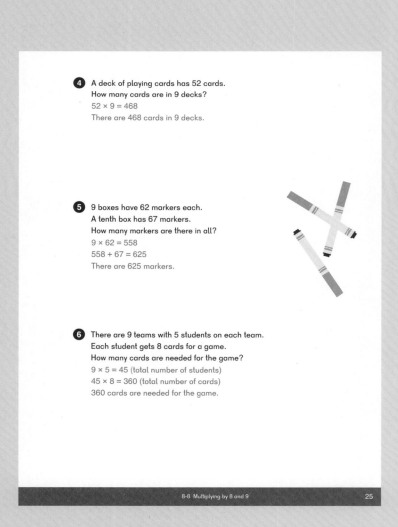

Exercise 9 • pages 26–28

Exercise 9

Basics

1. Divide 952 by 8.

   ```
       1 1 9
     8)9 5 2
       8
       1 5
         8
         7 2
         7 2
           0
   ```

   — 9 hundreds ÷ 8 is __1__ hundred with 1 hundred left over.

   — 15 tens ÷ 8 is __1__ ten with 7 tens left over.

   — 72 ones ÷ 8 is __9__ ones with 0 ones left over.

   952 ÷ 8 = __119__

   Check: __119__ × 8 = 952

2. Divide 672 by 9.

   ```
         7 4
     9)6 7 2
       6 3
         4 2
         3 6
           6
   ```

   672 ÷ 9 is __74__ with a remainder of __6__.

   Check: __74__ × 9 + 6 = 672

Practice

3. Divide.
   The sum of the remainders should equal the product of 9 and 2.

972 ÷ 8	473 ÷ 9	555 ÷ 9	683 ÷ 8
121 R 4	52 R 5	61 R 6	85 R 3

   4 + 5 + 6 + 3 = 18
   9 × 2 = 18

4. A rope is 458 m long.
   It is cut into pieces that are each 8 m long.
   How many pieces are there?
   How long is the leftover piece?
   458 ÷ 8 is 57 R 2
   There are 57 pieces.
   The left-over piece is 2 m long.

5. A baker made 250 cookies and gave away 34 of them.
   She put the rest equally into 9 tins.
   How many cookies are in each tin?
   250 − 34 = 216
   216 ÷ 9 = 24
   There are 24 cookies in each tin.

6. Find the missing numbers.
   What pattern do you notice?

   100 ÷ 9 is __11__ with a remainder of __1__.

   200 ÷ 9 is __22__ with a remainder of __2__.

   300 ÷ 9 is __33__ with a remainder of __3__.

   400 ÷ 9 is __44__ with a remainder of __4__.

   500 ÷ 9 is __55__ with a remainder of __5__.

   600 ÷ 9 is __66__ with a remainder of __6__.

   700 ÷ 9 is __77__ with a remainder of __7__.

   800 ÷ 9 is __88__ with a remainder of __8__.

   900 ÷ 9 is __100__ with a remainder of __0__.

   When hundreds are divided by 9 up to 800, the digits in the quotient and the remainder are the same as the digit in the hundreds place.

Exercise 10 • pages 29–32

Exercise 10

Check

1 The numbers 1 through 10 appear only once in each shaded row and column.
Complete the multiplication table.

×	7	2	8	6	3	10	5	9	1	4
3	21	6	24	18	**9**	30	15	27	3	12
6	42	12	48	**36**	18	60	30	54	6	24
1	7	2	8	6	3	10	5	9	**1**	4
5	35	10	40	30	15	50	**25**	45	5	20
2	14	**4**	16	12	6	20	10	18	2	8
7	**49**	14	56	42	21	70	35	63	7	28
4	28	8	32	24	12	40	20	36	4	**16**
9	63	18	72	54	27	90	45	**81**	9	36
10	70	20	80	60	30	**100**	50	90	10	40
8	56	16	**64**	48	24	80	40	72	8	32

2 Multiply or divide.

9 × 847 = 7,623 **L**	680 ÷ 8 = 85 **E**	216 ÷ 9 = 24 **R**
792 ÷ 8 = 99 **S**	61 × 9 = 549 **K**	8 × 792 = 6,336 **A**
8 × 38 = 304 **A**	495 ÷ 9 = 55 **W**	974 × 8 = 7,792 **H**

What is the largest type of shark?
Write the letters that match the answers above to find out.

W	H	A	L	E	S	H	A	R	K
55	7,792	304	7,623	85	76	99	7,792	6,336	24	549

3 Write the missing digits.

(a) 757 × 8 = 6,056

(b) 125 × 9 = 1,125

(c) 76 × 8 = 609 ... remainder 1

(d) 86 × 9 = 777 ... remainder 3

4 An orchard has 8 times as many apple trees as pear trees.
There are 981 trees in all.
How many more apple trees than pear trees does it have?

The orchard has 763 more apple trees.

9 units → 981
1 unit → 981 ÷ 9 = 109
7 units → 7 × 109 = 763

5 A box holds 8 large bowls and 6 small bowls.
Wanda needs 250 large bowls for a restaurant.
How many of these boxes does she need to buy?
How many small bowls will she also have?
250 ÷ 8 is 31 R 2
She needs to buy 32 boxes.
32 × 6 = 192
She will also have 192 small bowls.

6 (a) 78 = [9] × 8 + 6

(b) 47 = 7 × [6] + 5

7 9 children are playing a game with 126 counters.
Each child gets the same number of counters.
2 children left and the counters were shared out again.
How many more counters does each child get now than before?
126 ÷ 9 = 14 (number of counters per child first)
9 − 2 = 7
126 ÷ 7 = 18 (number of counters per child after 2 children left)
18 − 14 = 4
Each child has 4 more counters now than they did before.

Challenge

8 A park has 9 trees planted in a row,
with 7 bushes between each tree.
How many bushes are there?
There are 8 gaps total between the trees.
8 × 7 = 56
There are 56 bushes.

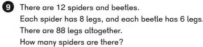

9 There are 12 spiders and beetles.
Each spider has 8 legs, and each beetle has 6 legs.
There are 88 legs altogether.
How many spiders are there?
If they are all beetles, that would be 12 × 6 = 72 legs.
88 − 72 = 16
2 more legs have to go on each spider.
16 ÷ 2 = 8
There are 8 spiders.
Check: Spider legs: 8 × 8 = 64; Beetles: 12 − 8 = 4;
Beetle legs: 4 × 6 = 24; Total legs: 24 + 64 = 88

Chapter 9 Fractions — Part 1

Overview

Suggested number of class periods: 5–6

	Lesson	Page	Resources	Objectives
	Chapter Opener	p. 45	TB: p. 33	Investigate fractions in the context of measurement.
1	Fractions of a Whole	p. 46	TB: p. 34 WB: p. 33	Understand that common fractions are composed of unit fractions. Find two fractions with the same denominator that make a whole.
2	Fractions on a Number Line	p. 49	TB: p. 39 WB: p. 37	Understand that fractions are numbers and can be represented on a number line.
3	Comparing Fractions with Like Denominators	p. 52	TB: p. 43 WB: p. 40	Compare fractions with the same denominator.
4	Comparing Fractions with Like Numerators	p. 55	TB: p. 47 WB: p. 43	Compare fractions with like numerators.
5	Practice	p. 58	TB: p. 51 WB: p. 47	Practice concepts from the chapter.
	Workbook Solutions	p. 60		

Chapter 9 Fractions — Part 1

Notes

In Dimensions Math 2B, students learned:

- The term "fractions" represents numbers that count equally divided parts of the whole.
- A unit fraction is a fraction with a numerator of 1.
- How fractions are composed using unit fractions.
- How to compose and decompose two fractions that make 1 whole.
- How to compare and order unit fractions.

In this chapter, students learn the terms "numerator," "denominator," and "unit fraction." Using an area model and number lines, students will use reasoning to compare and order fractions.

The Chapter Opener allows exploration of fractions in a real world context by using a meter as the whole unit. Students will look for objects in the classroom with lengths that can be expressed in fractional units.

Whole numbers can be represented as units of 1. By counting the units, we can express the sizes of numbers. For example, 5 can be represented as 5 units of 1.

Students will encounter fractions in different contexts:

- Two of three equally divided parts.
- The quantity resulting from a measurement, such as $\frac{2}{3}$ meter or $\frac{2}{3}$ deciliter.

These two contexts are addressed in this chapter.

In Dimensions Math 4 and 5, students will be introduced to:

- The ratio of A to B, i.e., the relative size of A compared to B.
- The quotient of 2 ÷ 3.

What is a Fraction?

Lesson 1 reviews the term "fractions" as numbers that count equally divided parts of the whole. The lesson also introduces the terms "numerator" and "denominator."

The denominator is the number of parts a whole is divided into. It names the fractional unit. The numerator is the number of those parts being counted. It counts the fractional units.

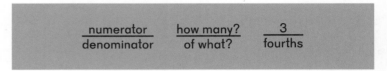

In the fraction $\frac{3}{4}$, one whole is divided into 4 parts. Each part is the unit fraction $\frac{1}{4}$. If we count 3 such parts, we count 3 one-fourths. That is, we have 3 fourths, which is written as $\frac{3}{4}$.

Area and linear models are used to show fractions in Lesson 1:

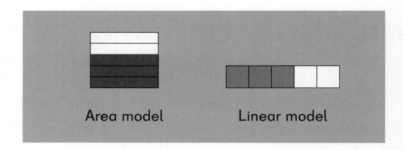

Area model   Linear model

Lesson 2 introduces a number line (linear) model and extends to fractions greater than 1.

To show the fraction $\frac{3}{4}$ on a number line, a segment from 0 to 1 is divided into 4 equal-length parts and the parts are labeled consecutively. The distance between each fraction on a number line is a unit fraction.

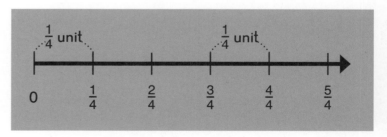

$\frac{3}{4}$ is 3 counts of $\frac{1}{4}$ unit.

A fraction is represented by counts of unit fractions. This helps students understand that a fraction could be greater than one whole. For example, 4 counts of one-fourth is 4 fourths ($\frac{4}{4}$), which is 1, and 5 counts of one-fourth is 5 fourths, which can be written as $\frac{5}{4}$.

Chapter 9 Fractions — Part 1

On a number line, students should see that $\frac{1}{4}$ is the same as $\frac{1}{4}$ of the distance from 0 to 1. $\frac{5}{6}$ inches can be thought of as $\frac{5}{6}$ inches away from 0. This concept will help students to draw the tick marks — four tick marks are needed between 0 and 1 to marks fifths, because the first tick mark is one-fifth away from 0, the second tick mark is 2 one-fifths away from 0, and so on.

Students also need to understand that a fraction is a number.

Comparing Fractions

In Lessons 3 and 4, students will compare fractions. Comparing fractions with like denominators helps students understand the meaning of the numerator (it counts the fractional units). Comparing fractions with like numerators helps students understand the meaning of the denominator (it names the fractional unit).

Comparing Fractions with Common Denominators

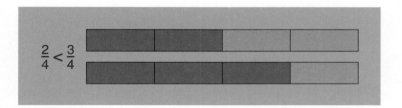

Because the numerator counts the number of parts, it is easy to compare fractions with the same denominator: Number lines also make it easy to compare fractions with the same denominator. A fraction farther to the right of 0 on a number line is greater than the fractions that come before it.

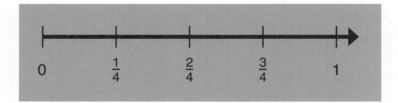

Comparing Fractions with Common Numerators

Comparing fractions with the same numerator can be confusing for students at first, however, with careful attention, students will master this important concept.

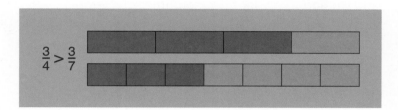

Students must understand that $\frac{1}{7}$ is smaller than $\frac{1}{4}$. If 1 is divided into 7 equal pieces, each part is smaller than if 1 was divided into 4 equal pieces.

Through recognizing that although the number in the denominator is greater, the size of the piece is smaller, students are able to develop critical understanding of denominators.

Students will compare fractions using other strategies in Chapter 10: Fractions — Part 2.

Chapter 9 Fractions — Part 1　　Materials

Materials

- 1-meter strips of paper or adding machine tape
- Different length strips of paper or adding machine tape
- Dry erase markers
- Fractions tiles for fifths, eighths, and tenths
- Paper strip or adding tape that is at least 12 in long
- String
- Whiteboards

Blackline Masters

- Comparing Fractions with Like Denominators
- Fraction Match Cards
- Order Up Cards

Activities

Activities included in this chapter are designed to provide practice with understanding and comparing. They can be used after students complete the **Do** questions, or any time additional practice is needed.

Chapter Opener

Objective

- Investigate fractions in the context of measurement.

Lesson Materials

- 1-meter strips of paper or adding machine tape

Provide students with 1-meter long strips of paper and have them measure something shorter than the strip.

Ask students:

- "How can you express the length if the 1-meter long strip is the only unit you have, i.e., no centimeters, meters, feet, or inches?"
- "How can you express the length so someone else has a rough idea of the length if we could not use centimeters, meters, feet, or inches?"

Encourage students to think about parts of a whole. Remind them of what they learned in previous grades.

Students should see that they can fold the paper strip into equal parts to help them measure something less than the length of the paper strip.

Discuss the illustration on Textbook page 33.

Have students compare the items they measured with the items the friends are measuring.

This informal investigation allows students to see fractions in real life. In this lesson, our unit is 1 whole.

Lesson 1 Fractions of a Whole

Objectives

- Understand that common fractions are composed of unit fractions.
- Find two fractions with the same denominator that make a whole.

Lesson Materials

- Different length strips of paper or adding machine tape

Think

Provide students with paper and pose the **Think** problems. Have students fold their paper into thirds.

Have students try to solve the problem on their own and then share their solutions. Students should see that although the paper strips are different lengths, each part can be considered as $\frac{1}{3}$ of the whole strip.

Learn

Sofia, Dion, and Emma show how to write each fraction. Students should see that $\frac{2}{3}$ is 2 one-thirds.

Discuss the terms "fraction," "numerator," and "denominator." Regardless of the size of the whole, the denominator shows how many parts there are in that whole and the numerator counts the number of parts in that whole.

While the initial concepts in **Think** are review from Dimensions Math 2B, the terminology in **Learn** is new.

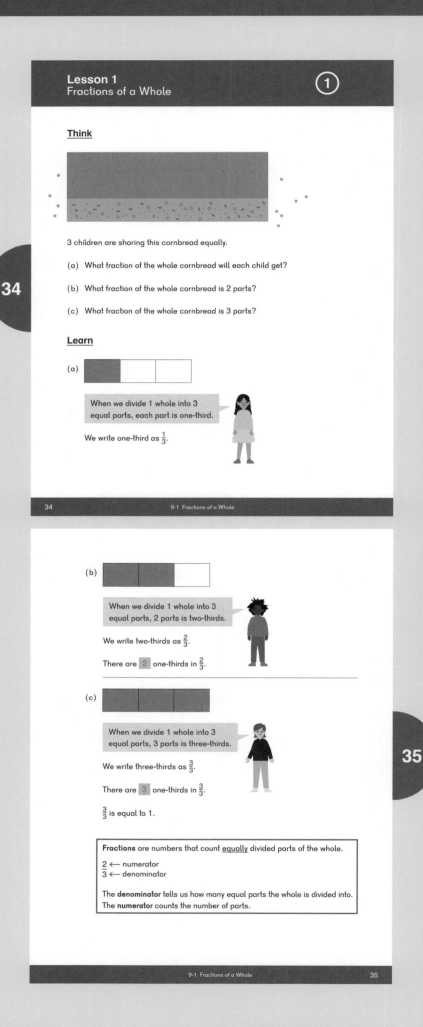

46   Teacher's Guide 3B Chapter 9   © 2017 Singapore Math Inc.

Do

2 – 3 These problems review finding the fraction that makes a whole with a given fraction.

2 Mei remembers that 1 whole is $\frac{4}{4}$. Expand on this to ensure students realize that the whole can be divided into many parts and all of those parts make a whole. For example, $\frac{5}{5}$, $\frac{9}{9}$, and $\frac{100}{100}$ are all equal to 1.

3 Students should see that the shaded parts need not be contiguous to show a fraction, that is, $\frac{5}{7}$ is any 5 parts of a whole that is divided into 7 equally-sized parts.

4 In each of these problems, the whole is the whole shape.

6 Here, the whole is $\frac{5}{5}$ or 1 liter of water. There is $\frac{3}{5}$ L of water in the beaker.

(b) $\frac{3}{5} = \frac{1}{5}$ and $\frac{1}{5}$ and $\frac{1}{5}$, or, $\frac{3}{5}$ is 3 fifths or 3 one-fifths.

7 Emma defines a unit fraction. Ensure that students understand that a fraction can be composed of unit fractions.

(a) 10 units of $\frac{1}{10}$ m make 1 whole meter.

8 Students can recall their paper strips from **Think** to see that $\frac{1}{3}$ of different-sized wholes are not the same. In this problem, $\frac{2}{3}$ of a meter is not the same length as $\frac{2}{3}$ of a foot.

Activities

● **Match**

Materials: Fraction Match Cards (BLM)

Use this game as a review of fractions that make one whole.

Lay cards in a faceup array. Have students match two Fraction Match Cards (BLM) with fractions that, when combined, make one whole.

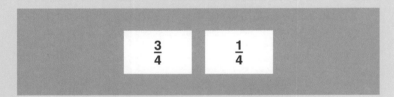

▲ **Memory**

Materials: Fraction Match Cards (BLM)

Play using the same rules as **Match**, but set the cards out facedown in an array.

▲ **Show Me**

Materials: Whiteboards, dry erase markers

Tell students we are making 1. Give them a part and have them show you a part that adds to yours to make 1. For example, "We are finding fractions that make 1. My part is $\frac{3}{5}$. What is your part?"

Exercise 1 • page 33

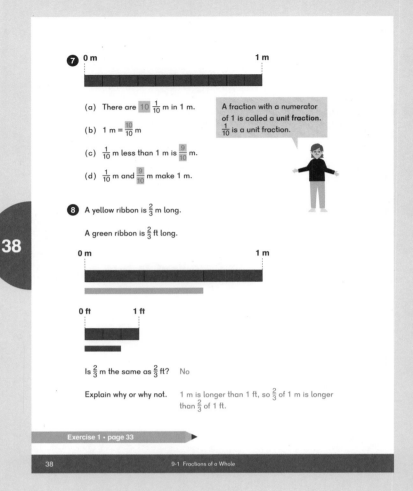

Lesson 2 Fractions on a Number Line

Objective

- Understand that fractions are a number and can be represented on a number line.

Lesson Materials

- Paper strip or adding tape that is at least 12 in long

Think

Provide students with a strip of paper. Have them fold the strip into four equal parts, similar to Dion's strip in the Chapter Opener on Textbook page 33.

Have students repeat the Chapter Opener task of finding objects in the classroom that are longer than the paper strip. Discuss how they would name that length using fractions.

Pose the **Think** problems and have students try to answer the questions independently. They should note that the length of the red ribbon in **Think** is longer than 1 meter.

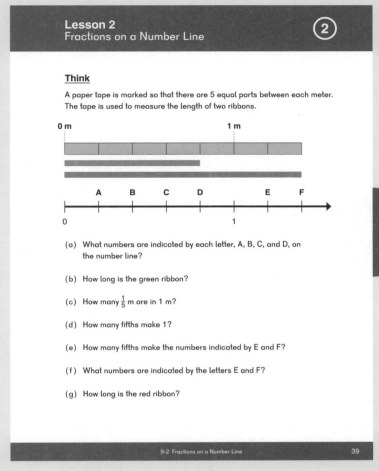

Learn

Discuss and then demonstrate Alex's comment:

- Draw a number line on the front board. Label the left end 0 and a point on the line as 1.
- Using tick marks, divide the distance between 0 and 1 into 5 equal parts.
- Have students say the name of the parts. (fifths)
- Ask them what to label the first tick mark ($\frac{1}{5}$), the second tick mark ($\frac{2}{5}$), up to the one whole ($\frac{5}{5}$ and 1).
- Add a tick mark to represent $\frac{6}{5}$ and ask students what to label that tick mark.

Ensure that students understand that the tick marks on the number line show the unit fractions. Help students to see that 4 tick marks between 0 and 1 make 5 equal parts. The distance between $\frac{3}{5}$ and $\frac{4}{5}$ is $\frac{1}{5}$ of the whole.

Students have previously thought of fractions as parts of a whole. Sofia tells us that fractions are also numbers on a number line.

Repeat with other number lines showing thirds, fourths, sixths, or tenths as needed.

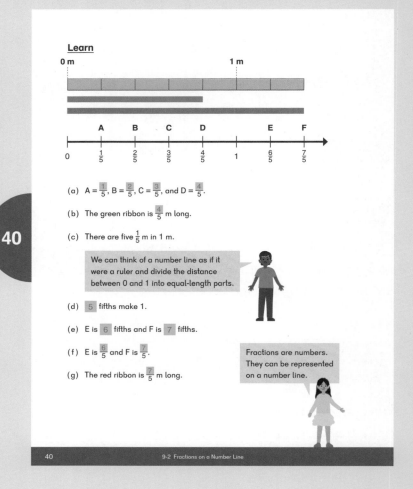

50　　　　　　　　　　　Teacher's Guide 3B Chapter 9　　　　　　　　　© 2017 Singapore Math Inc.

Do

1. In **Think** and **Learn**, the number line represented 1 meter. Here the number line simply measures the number of fractional units in 1.

   (c) and (d) are laying the foundation for adding and subtracting fractions.

2. — 3. These problems extend the fractional units on a number line beyond 1 whole.

4. — 5. Have students compare the ruler images in the book to real rulers.

   In 4, students may see the 2-inch mark as $\frac{8}{4}$, the 3-inch mark as $\frac{12}{4}$, etc.

Activity

▲ **Choral Counting**

Using your thumb pointing up or down, have students chorally count on and back by thirds, fourths, etc.

Example: You say, "Let's count by $\frac{1}{3}$ starting at 0. First number?" Class responds, "0." Point thumb up, class responds, "$\frac{1}{3}$." Point thumb up again. Class responds, "$\frac{2}{3}$." Point down. Class responds, "$\frac{1}{3}$."

Start at random fractions. You say, "Let's count by fourths starting at $\frac{3}{4}$. First number?" Class responds, "$\frac{3}{4}$." Point thumb up. Class responds, "$\frac{4}{4}$." Point thumb up again. Class responds, "$\frac{5}{4}$." Point thumb down. Class responds, "$\frac{4}{4}$."

Exercise 2 • page 37

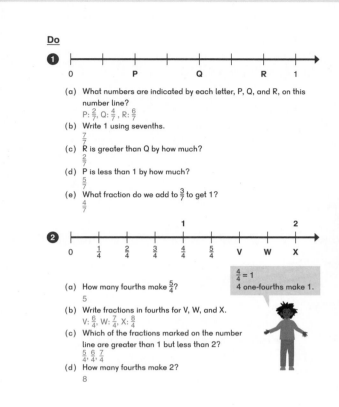

© 2017 Singapore Math Inc.   Teacher's Guide 3B Chapter 9   51

Lesson 3 Comparing Fractions with Like Denominators

Objective
- Compare fractions with the same denominator.

Lesson Materials
- Comparing Fractions with Like Denominators (BLM)

Think

Provide students with Comparing Fractions with Like Denominators (BLM) and pose the **Think** problem. Students can label, shade, and cut out each model, or draw their own.

Discuss student strategies for comparing the amount of paint the friends have. Have students write the fractions in order from least to greatest.

Students should see that the denominators are the same.

Learn

Have students compare their solutions from **Think** with the one shown in the textbook.

Discuss Mei and Emma's comments.

Dion points out that it is easy to compare fractions with the same denominator on a number line. As we count from left to right on a number line, the numbers on the right are greater than the numbers on the left.

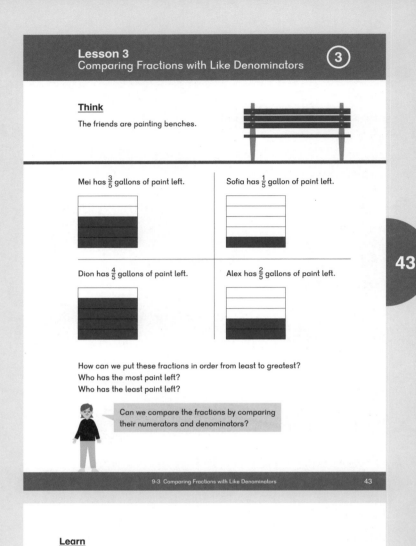

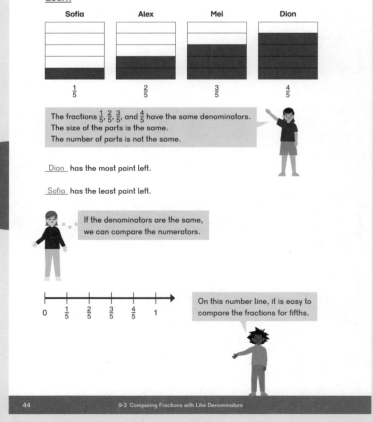

52 Teacher's Guide 3B Chapter 9 © 2017 Singapore Math Inc.

Do

Ensure students understand that we can only compare fractions that have the same whole. Pizza can be a relatable example for this.

Have students imagine that Alex went to PizzaPalooza and bought a large pizza. He ate $\frac{1}{5}$ of his pizza.

Emma went to Pizza Shoppe and bought a large pizza. She also ate $\frac{1}{5}$ of her pizza.

Show students an example similar to the one below and ask them: "Can we say they each ate the same amount of pizza? Why or why not?"

Size Large from PizzaPalooza   Size Large from Pizza Shoppe

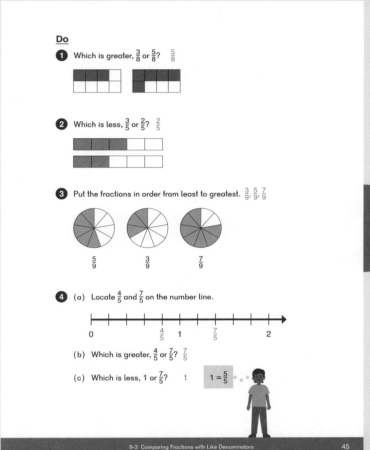

Note the progression of representations in the questions:

❶ — ❸ include a picture.

❹ is slightly more abstract, showing a number line.

❺ — ❾ provide the abstract fractions with numerals only.

5 – 7 By this point, students should understand that when the denominators are equal, they can simply count to find which fraction has the greater numerator (or more of those same-sized parts).

8 (d) Encourage students to draw a number line if they are confused.

9 (b) Ask students how they solved this problem. Some may think of 1 L as $\frac{10}{10}$ L and know that is more than $\frac{7}{10}$ L.

Activity

▲ Fraction Battle – Common Denominator

Materials: Fraction Match Cards (BLM), sorted into sets of common denominators for each group (for example, $\frac{1}{5}$, $\frac{2}{5}$, $\frac{3}{5}$, and $\frac{4}{5}$)

Playing in a group of 2 to 3 students, evenly deal out all Fraction Match Cards (BLM) facedown.

Players each flip over a card at the same time. The greatest fraction (or least, depending on version of game) wins.

Exercise 3 • page 40

---

5 Which is greater, $\frac{3}{7}$ or $\frac{5}{7}$? $\frac{5}{7}$

6 Which is less, $\frac{17}{12}$ or $\frac{11}{12}$? $\frac{11}{12}$

7 What sign, >, <, or =, goes in the ◯?

(a) $\frac{1}{4}$ ◯ $\frac{3}{4}$ (b) $\frac{8}{8}$ ◯ 1

(c) $\frac{3}{4}$ ◯ $\frac{6}{4}$ (d) $\frac{6}{9}$ ◯ $\frac{5}{9}$

(e) $\frac{10}{8}$ ◯ $\frac{3}{8}$ (f) $\frac{7}{16}$ ◯ $\frac{5}{16}$

8 Put the numbers in order from least to greatest.

(a) 1, 0, $\frac{1}{2}$ 0, $\frac{1}{2}$, 1   (b) $\frac{5}{8}$, $\frac{3}{8}$, $\frac{6}{8}$ $\frac{3}{8}$, $\frac{5}{8}$, $\frac{6}{8}$

(c) $\frac{2}{6}$, $\frac{5}{6}$, $\frac{8}{6}$, $\frac{3}{6}$ $\frac{2}{6}$, $\frac{3}{6}$, $\frac{5}{6}$, $\frac{8}{6}$ (d) $\frac{7}{8}$, 2, $\frac{5}{8}$, 1, $\frac{9}{8}$ $\frac{5}{8}$, $\frac{7}{8}$, 1, $\frac{9}{8}$, 2

9 Wainani drank $\frac{7}{10}$ L of water at school today. Olivia drank 1 L of water.

(a) Who drank more water?
    Olivia
(b) How much more water?
    $1 - \frac{7}{10} = \frac{3}{10}$
    $\frac{3}{10}$ L more

Exercise 3 • page 40

46   9-3 Comparing Fractions with Like Denominators

Lesson 4 Comparing Fractions with Like Numerators

Objective

- Compare fractions with like numerators.

Lesson Materials

- Fractions tiles for fifths, eighths, and tenths

Think

Discuss the **Think** problem and Sofia's comment. Ask students how this problem is different from the previous lesson.

- The friends have water, not paint.
- Each friends' parts are different sizes.
- The numerators are the same, but the denominators are different.

Provide students with fraction tiles and time to investigate with the tiles. Then have students find solutions to the **Think** questions. Remind students that 1 whole represents 1 liter of water.

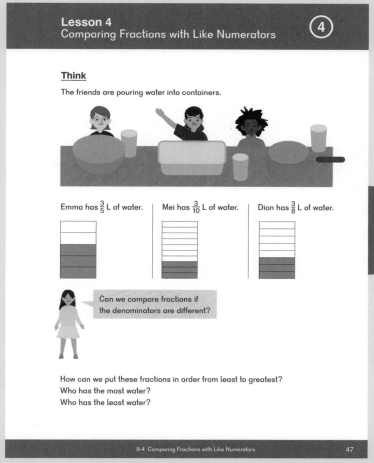

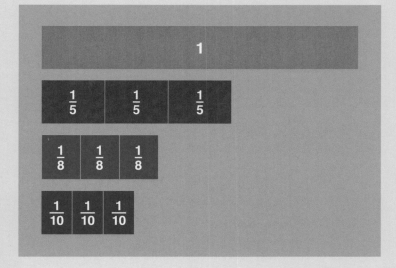

Discuss how students compared and ordered the fractions. Ask them how they would compare the fractions if they did not have tiles or the picture.

Learn

Mei points out a key understanding of fractions. As the value in the denominator increases, the size of each unit fraction is smaller. Dion expands on this idea by showing the unit fractions on a number line.

Have students draw two number lines of the same length. Then have them divide the first number line into 5 equal parts and the second into 10 equal parts. If $\frac{1}{5}$ is greater than $\frac{1}{10}$, then 3 of those fifths is greater than 3 of the tenths.

Do

❷ Alex thinks of comparing the unit fractions. He knows that if 1 ninth is less than 1 seventh, then 5 ninths will be less than 5 sevenths.

Note the progression of representations in the questions:

❶—❷ include a picture.

❸ is slightly more abstract, showing a number line.

❹—❽ provide the abstract fractions with numerals only.

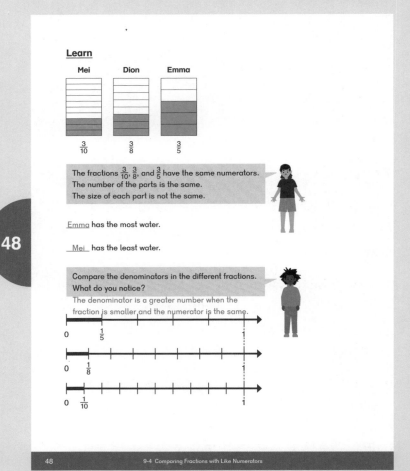

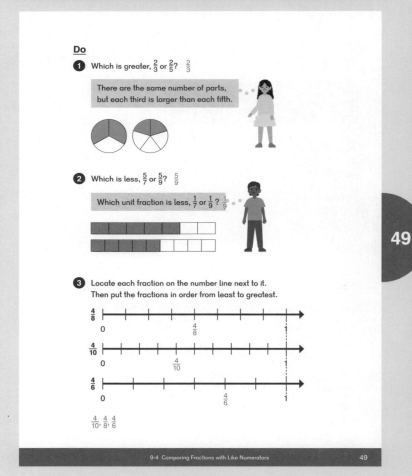

7 (d) Ask students:

- Are the numerators all the same?
- What happens with 0 and 1?
- How do we know 1 is greatest?

Activity

▲ Fraction Battle – Common Numerator

Materials: Fraction Match Cards (BLM), sorted into sets of common numerators for each group (for example: $\frac{5}{6}, \frac{5}{7}, \frac{5}{8}, \frac{5}{9}, \frac{5}{10}$)

Playing in a group of 2 to 3 students, evenly deal out all Fraction Match Cards (BLM) facedown.

Players each flip over a card at the same time. The greatest fraction (or least, depending on version of game) wins.

Exercise 4 • page 43

---

4 Which is greater, $\frac{3}{4}$ or $\frac{3}{5}$? $\frac{3}{4}$

5 Which is less, $\frac{11}{12}$ or $\frac{11}{16}$? $\frac{11}{16}$

6 What sign, >, <, or =, goes in the ◯?

(a) $\frac{1}{8} \bigcirc< \frac{1}{3}$ (b) $\frac{2}{3} \bigcirc> \frac{2}{5}$

(c) $\frac{6}{9} \bigcirc< \frac{6}{8}$ (d) $\frac{5}{7} \bigcirc> \frac{5}{8}$

(e) $\frac{10}{8} \bigcirc< \frac{10}{7}$ (f) $\frac{9}{12} \bigcirc< \frac{9}{10}$

7 Put the numbers in order from least to greatest.

(a) $\frac{1}{3}, \frac{1}{8}, \frac{1}{6}$ $\frac{1}{8}, \frac{1}{6}, \frac{1}{3}$ (b) $\frac{7}{5}, \frac{7}{8}, \frac{7}{6}$ $\frac{7}{8}, \frac{7}{6}, \frac{7}{5}$

(c) $\frac{3}{4}, \frac{3}{7}, \frac{3}{5}, \frac{3}{8}$ $\frac{3}{8}, \frac{3}{7}, \frac{3}{5}, \frac{3}{4}$ (d) $\frac{2}{5}, 0, \frac{2}{8}, 1$ $0, \frac{2}{8}, \frac{2}{5}, 1$

8 A pine seedling is $\frac{7}{10}$ m tall.

A birch seedling is $\frac{7}{8}$ m tall.

Which tree seedling is taller?
The birch seedling is taller.

Exercise 4 • page 43

50   9-4 Comparing Fractions with Like Numerators

Lesson 5 Practice

Objective
- Practice concepts from the chapter.

After students complete **Practice** in the textbook, have them continue to practice ordering and comparing fractions with activities from the chapter.

Activity

▲ **Number Line**

Materials: Order Up Cards (BLM), string

Tie up the string or have students hold it up at either end. Hand out the 0 and 1 cards to students and have them place the cards on the string number line.

Hand out the remaining Order Up Cards (BLM) and have students add their cards to the string number line, one at a time. They will need to adjust the other cards to place their cards into the appropriate place:

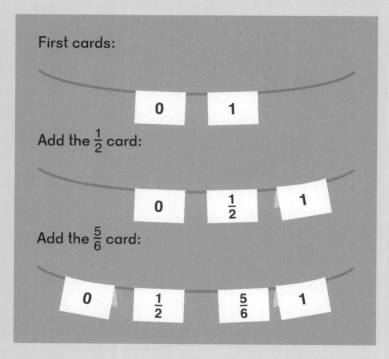

Students can reason that $\frac{5}{6}$ is almost $\frac{6}{6}$, and belongs closer to 1 than to $\frac{1}{2}$.

If students struggle, have them discuss where each new card should be placed.

7. Have students make up similar puzzles to challenge classmates.

Brain Works

★ Fractions on a Number Line

Create 4 fractions, using each number, 1 through 9, only once (you will only be able to use eight of the numbers each time) as either numerator or denominator. Place them all on the number line from least to greatest.

How many solutions can you find?

Example answers:

$\frac{1}{9}, \frac{2}{7}, \frac{3}{4}, \frac{6}{5}$

Reasoning:

$\frac{1}{9}$ = One small part

$\frac{2}{7}$ = two larger parts than ninths

$\frac{3}{4}$ = more than one-half

$\frac{6}{5}$ = more than 1

Exercise 5 • page 47

---

4. Which of these fractions are greater than 1?

$\frac{5}{7}$ $\boxed{\frac{5}{3}}$ $\frac{6}{6}$ $\frac{7}{8}$ $\boxed{\frac{9}{7}}$

5. What sign, >, <, or =, goes in the ◯?

(a) $\frac{3}{7} \boxed{<} \frac{3}{4}$ (b) $\frac{5}{8} \boxed{>} \frac{3}{8}$

(c) $\frac{6}{6} \boxed{>} \frac{1}{3}$ (d) $\frac{5}{3} \boxed{>} \frac{2}{3}$

(e) $\frac{5}{5} \boxed{=} \frac{10}{10}$ (f) $\frac{7}{4} \boxed{>} \frac{4}{7}$

6. Put the numbers in order from least to greatest.

(a) $\frac{2}{7}, \frac{5}{7}, \frac{7}{3}$ $\frac{2}{7}, \frac{5}{7}, \frac{7}{3}$

(b) $\frac{5}{7}, \frac{5}{9}, \frac{5}{12}$ $\frac{5}{12}, \frac{5}{9}, \frac{5}{7}$

(c) $\frac{5}{3}, 1, \frac{1}{3}, 0$ $0, \frac{1}{3}, 1, \frac{5}{3}$

(d) $\frac{7}{5}, \frac{7}{8}, \frac{4}{8}, \frac{4}{9}$ $\frac{4}{9}, \frac{4}{8}, \frac{7}{8}, \frac{7}{5}$

7. What fraction has a numerator of 3, is greater than $\frac{3}{5}$, and is less than 1?

$\frac{3}{4}$

8. Papina has done $\frac{2}{3}$ of her math problems. What fraction of her math problems does she have left to do?

$\frac{1}{3}$

Exercise 5 • page 47

Exercise 1 • pages 33–36

Chapter 9 Fractions — Part 1

Exercise 1

Basics

1. A pentagon is divided into 5 equal parts.
   3 parts are shaded.
   $\frac{3}{5}$ of the pentagon is shaded.

   (a) $\frac{3}{5}$ is __3__ out of __5__ equal parts.

   (b) In $\frac{3}{5}$, the numerator is __3__ and the denominator is __5__.

   (c) One part is $\boxed{\frac{1}{5}}$ of the whole.

   (d) $\frac{3}{5}$ = __3__ fifths

   (e) 1 whole = __5__ fifths

   (f) $\frac{3}{5}$ and $\boxed{\frac{2}{5}}$ make 1 whole.

2.

   (a) On the bar, __7__ out of __10__ equal parts are shaded.

   (b) $\boxed{\frac{7}{10}}$ of the bar is shaded.

   (c) 1 whole = __10__ tenths

   (d) $\boxed{\frac{7}{10}}$ and $\boxed{\frac{3}{10}}$ make $\boxed{\frac{10}{10}}$, which is 1 whole.

9-1 Fractions of a Whole — 33

Practice

3. What fraction of each shape is shaded?

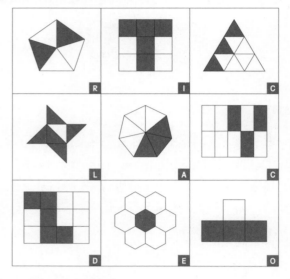

R   I   C
L   A   C
D   E   O

What animal cannot move its tongue?
Write the letters that match the answers above to find out.

A	C	R	O	C	O	D	I	L	E
$\frac{7}{12}$	$\frac{3}{7}$	$\frac{5}{8}$	$\frac{3}{9}$	$\frac{2}{5}$	$\frac{3}{4}$	$\frac{3}{10}$	$\frac{3}{4}$	$\frac{5}{12}$	$\frac{5}{9}$	$\frac{5}{6}$	$\frac{1}{7}$	$\frac{1}{6}$

34 — *9-1 Fractions of a Whole*

4. Shade the given fraction for each shape.
   Shaded sections of each shape may vary.

   (a) $\frac{6}{8}$ (b) $\frac{5}{10}$

   (c) $\frac{4}{12}$ (d) $\frac{4}{9}$

   (e) $\frac{6}{9}$ (f) $\frac{5}{8}$

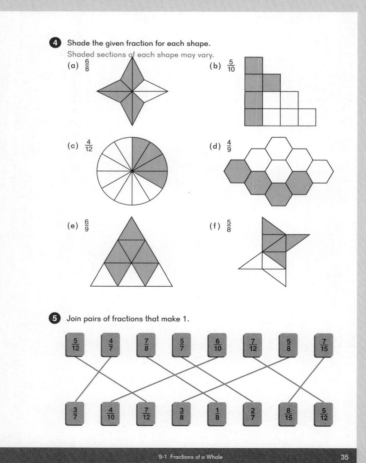

5. Join pairs of fractions that make 1.

   $\frac{5}{12}$ $\frac{4}{7}$ $\frac{7}{8}$ $\frac{5}{7}$ $\frac{6}{10}$ $\frac{7}{12}$ $\frac{5}{8}$ $\frac{7}{15}$

   $\frac{3}{7}$ $\frac{4}{10}$ $\frac{7}{12}$ $\frac{3}{8}$ $\frac{1}{8}$ $\frac{2}{7}$ $\frac{8}{15}$ $\frac{5}{12}$

9-1 Fractions of a Whole — 35

Challenge

6. Use a straightedge to split each shape below into the number of equal parts given in the denominator of each fraction.
   Then shade the number of parts given in the numerator of each fraction.
   Shaded sections of each shape may vary.

   (a) $\frac{4}{5}$ (b) $\frac{1}{2}$

   (c) $\frac{7}{12}$ (d) $\frac{5}{7}$

   (e) $\frac{3}{4}$ (f) $\frac{2}{3}$

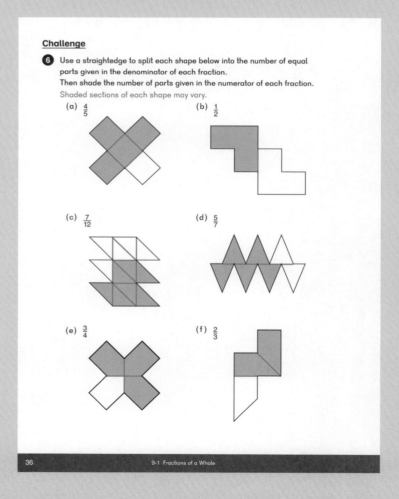

36 — *9-1 Fractions of a Whole*

Exercise 2 • pages 37–39

Exercise 2

Basics

1. A one-meter tape has been divided into 5 equal parts. 3 parts are shaded.

   (a) $\frac{3}{5}$ of the tape is shaded.

   (b) The shaded part is $\frac{3}{5}$ of a meter long.

   (c) The rope is $\frac{3}{5}$ m long.

2. Answer the questions based on the number line.

   (a) The number line is divided into __5__ equal parts between 0 and 1.

   (b) There are __5__ equal increments of $\frac{1}{5}$ between 0 and 1.

   (c) $\frac{5}{5}$ is the same as __1__.

   (d) There are __3__ equal increments of $\frac{1}{5}$ between 0 and $\frac{3}{5}$.

   (e) There are 2 equal increments between $\frac{3}{5}$ and 1.

   (f) $\frac{5}{5}$ is $\frac{1}{5}$ and $\frac{4}{5}$.

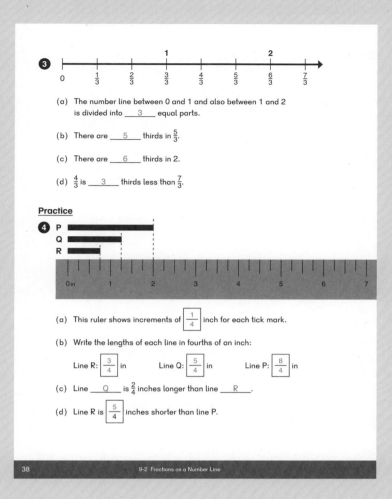

3.
   (a) The number line between 0 and 1 and also between 1 and 2 is divided into __3__ equal parts.

   (b) There are __5__ thirds in $\frac{5}{3}$.

   (c) There are __6__ thirds in 2.

   (d) $\frac{4}{3}$ is __3__ thirds less than $\frac{7}{3}$.

Practice

4.
   (a) This ruler shows increments of $\frac{1}{4}$ inch for each tick mark.

   (b) Write the lengths of each line in fourths of an inch:

   Line R: $\frac{3}{4}$ in   Line Q: $\frac{5}{4}$ in   Line P: $\frac{8}{4}$ in

   (c) Line __Q__ is $\frac{2}{4}$ inches longer than line __R__.

   (d) Line R is $\frac{5}{4}$ inches shorter than line P.

5. Label the numbers marked with arrows on each number line. Use those fractions to answer the questions below.

   (a) $\frac{1}{10}$, $\frac{8}{10}$, $\frac{11}{10}$

   (b) $\frac{2}{7}$, $\frac{5}{7}$, $\frac{9}{7}$

   (c) $\frac{5}{12}$, $\frac{11}{12}$, $\frac{15}{12}$

   (d) $\frac{1}{6}$, $\frac{5}{6}$, $\frac{11}{6}$, $\frac{15}{6}$

   (e) List the fractions from problems (a) through (d) that are less than 1.
   $\frac{1}{10}, \frac{8}{10}, \frac{2}{7}, \frac{5}{7}, \frac{5}{12}, \frac{11}{12}, \frac{1}{6}, \frac{5}{6}$

   (f) List the fractions from problems (a) through (d) that are between 1 and 2.
   $\frac{11}{10}, \frac{9}{7}, \frac{15}{12}, \frac{11}{6}$

Challenge

6. (a) __25__ fifths make 5.   (b) __1,000__ tenths make 100.

   (c) $\frac{27}{3}$ make 9.   (d) $\frac{12}{4}$ make 3.

Exercise 3 • pages 40–42

Exercise 3

Basics

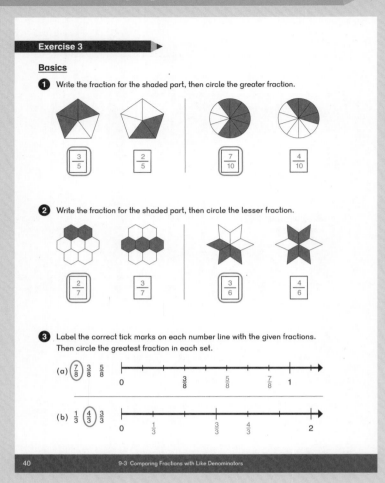

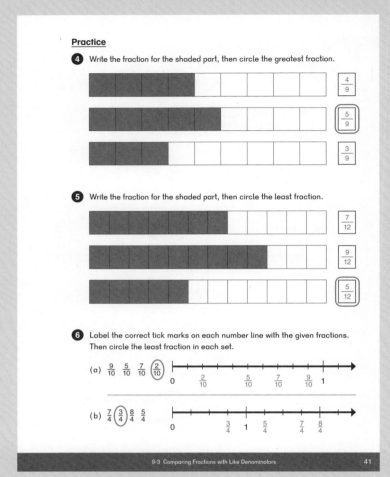

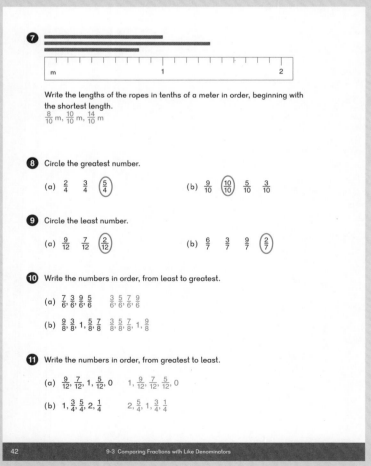

Exercise 4 • pages 43–46

Exercise 4

Basics

1. Write the fraction for the shaded part, then circle the greater fraction.

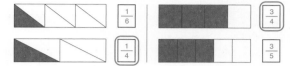

2. Write the fraction for the shaded part, then circle the lesser fraction.

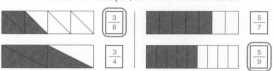

3. This number line shows 2 sets of tick marks.
   (a) Label the tick marks marked with arrows.

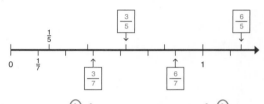

   (b) Which is greater, $\frac{3}{5}$ or $\frac{3}{7}$?   (c) Which is less, $\frac{6}{5}$ or $\frac{6}{7}$?

Practice

4.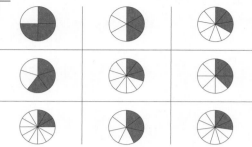

Write the fractions for the shaded parts above in order from least to greatest.

Least — Greatest

| $\frac{3}{12}$ | $\frac{3}{11}$ | $\frac{3}{10}$ | $\frac{3}{9}$ | $\frac{3}{8}$ | $\frac{3}{7}$ | $\frac{3}{6}$ | $\frac{3}{5}$ | $\frac{3}{4}$ |
| Blue | Red | Yellow | White | Black | Blue | | Black | Red | White |

Color the flags using the color code for each fraction above to see how to signal "I LOVE MATH" using maritime flags.

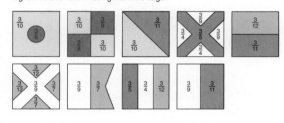

5. Color the bars to show the given fraction. Then write > or < in each ○.

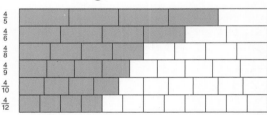

   (a) $\frac{4}{8} < \frac{4}{5}$ (b) $\frac{4}{10} < \frac{4}{6}$ (c) $\frac{4}{9} > \frac{4}{12}$

6. Label the correct tick marks on the correct number lines with the fractions listed below.
   Then write the fractions in order from least to greatest.

   $\frac{8}{9}, \frac{8}{12}, \frac{8}{7}, \frac{8}{10}$ $\frac{8}{12}, \frac{8}{10}, \frac{8}{9}, \frac{8}{7}$

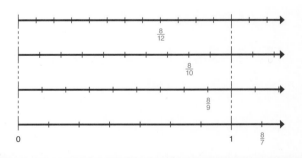

7. Circle the greatest number.
   (a) $\frac{1}{5}$ $\frac{1}{8}$ $\left(\frac{1}{4}\right)$
   (b) $\frac{7}{10}$ $\frac{7}{12}$ $\frac{7}{8}$ $\left(\frac{7}{4}\right)$

8. Circle the least number.
   (a) $\left(\frac{2}{8}\right)$ $\frac{2}{2}$ $\frac{2}{5}$
   (b) $\left(\frac{9}{13}\right)$ $\frac{9}{10}$ $\frac{9}{12}$ $\frac{9}{11}$

9. Write the numbers in order from least to greatest.
   (a) $\frac{1}{5}, \frac{1}{2}, \frac{1}{3}$ $\frac{1}{5}, \frac{1}{3}, \frac{1}{2}$
   (b) $\frac{5}{10}, \frac{5}{5}, \frac{5}{7}, \frac{5}{9}$ $\frac{5}{10}, \frac{5}{9}, \frac{5}{7}, \frac{5}{5}$

10. Write the numbers in order from greatest to least.
    (a) $\frac{5}{12}, \frac{5}{5}, 0, \frac{5}{9}$ $\frac{5}{5}, \frac{5}{9}, \frac{5}{12}, 0$
    (b) $\frac{2}{10}, \frac{2}{5}, \frac{2}{15}, \frac{2}{9}, \frac{2}{7}$ $\frac{2}{5}, \frac{2}{7}, \frac{2}{9}, \frac{2}{10}, \frac{2}{15}$

11. Write a numerator or denominator that will make each of the following true.
    Answers may vary. Possible solutions provided.
    (a) $\frac{5}{8} < \frac{5}{7}$ (b) $\frac{3}{10} > \frac{3}{12}$

Exercise 5 • pages 47–50

Exercise 5

Check

1 Multiply or divide.

68 × 8	365 × 4	589 × 7
544	1,460	4,123
97 ÷ 3	786 ÷ 5	634 ÷ 9
32 R 1	157 R 1	70 R 4

2 Which triangle correctly shows $\frac{3}{4}$ shaded? C
Explain why the other triangles are not $\frac{3}{4}$ shaded.

A — Not divided into equal parts.
B — Not divided into equal parts.
C — This is correct.
D — Divided into sixths, not fourths.

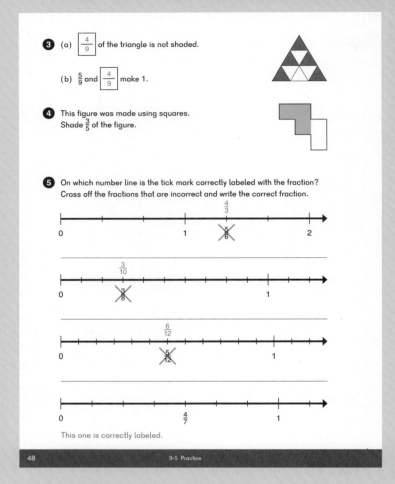

3 (a) $\boxed{\frac{4}{9}}$ of the triangle is not shaded.

(b) $\frac{5}{9}$ and $\boxed{\frac{4}{9}}$ make 1.

4 This figure was made using squares. Shade $\frac{3}{5}$ of the figure.

5 On which number line is the tick mark correctly labeled with the fraction? Cross off the fractions that are incorrect and write the correct fraction.

(Number line 1: $\frac{4}{3}$ correct, $\frac{5}{6}$ crossed off)
(Number line 2: $\frac{3}{10}$ correct, $\frac{3}{8}$ crossed off)
(Number line 3: $\frac{6}{12}$ correct, $\frac{5}{12}$ crossed off)
(Number line 4: $\frac{4}{7}$ — This one is correctly labeled.)

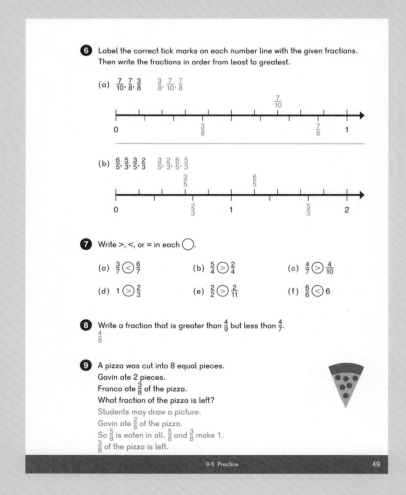

6 Label the correct tick marks on each number line with the given fractions. Then write the fractions in order from least to greatest.

(a) $\frac{7}{10}, \frac{7}{8}, \frac{3}{8}$ $\frac{3}{8}, \frac{7}{10}, \frac{7}{8}$

(b) $\frac{6}{5}, \frac{5}{3}, \frac{3}{5}, \frac{2}{3}$ $\frac{3}{5}, \frac{2}{3}, \frac{6}{5}, \frac{5}{3}$

7 Write >, <, or = in each ◯.

(a) $\frac{3}{7} < \frac{6}{7}$ (b) $\frac{5}{4} > \frac{2}{4}$ (c) $\frac{4}{7} > \frac{4}{10}$

(d) $1 > \frac{2}{3}$ (e) $\frac{2}{2} > \frac{2}{11}$ (f) $\frac{6}{6} < 6$

8 Write a fraction that is greater than $\frac{4}{9}$ but less than $\frac{4}{7}$.
$\frac{4}{8}$

9 A pizza was cut into 8 equal pieces.
Gavin ate 2 pieces.
Franco ate $\frac{3}{8}$ of the pizza.
What fraction of the pizza is left?
Students may draw a picture.
Gavin ate $\frac{2}{8}$ of the pizza.
So $\frac{5}{8}$ is eaten in all. $\frac{5}{8}$ and $\frac{3}{8}$ make 1.
$\frac{3}{8}$ of the pizza is left.

Challenge

10 Write the numbers in order, from least to greatest.

(a) $\frac{7}{6}, \frac{3}{6}, \frac{3}{7}, \frac{2}{7}, \frac{5}{6}$ $\frac{2}{7}, \frac{3}{7}, \frac{3}{6}, \frac{5}{6}, \frac{7}{6}$

(b) $\frac{9}{12}, \frac{12}{8}, 1, \frac{5}{12}, \frac{9}{8}$ $\frac{5}{12}, \frac{9}{12}, 1, \frac{9}{8}, \frac{12}{8}$

(c) $\frac{1}{5}, \frac{3}{3}, 3, \frac{8}{4}, \frac{1}{2}$ $\frac{1}{5}, \frac{1}{2}, \frac{3}{3}, \frac{8}{4}, 3$

11 Match each fraction to the bar that is shaded by that amount.

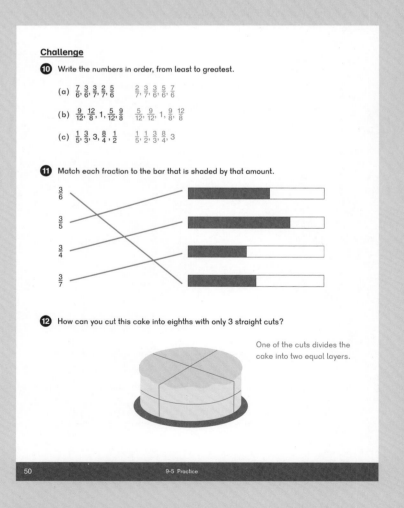

12 How can you cut this cake into eighths with only 3 straight cuts?

One of the cuts divides the cake into two equal layers.

Chapter 10 Fractions — Part 2

Overview

Suggested number of class periods: 9–10

	Lesson	Page	Resources	Objectives
	Chapter Opener	p. 69	TB: p. 53	Investigate equivalent fractions.
1	Equivalent Fractions	p. 70	TB: p. 54 WB: p. 51	Understand the meaning of equivalent fractions. Identify equivalent fractions using bar models or number lines.
2	Finding Equivalent Fractions	p. 72	TB: p. 58 WB: p. 55	Find equivalent fractions by multiplying the numerator and denominator by the same number.
3	Simplifying Fractions	p. 75	TB: p. 61 WB: p. 58	Find equivalent fractions by dividing the numerator and denominator by the same number. Simplify fractions.
4	Comparing Fractions — Part 1	p. 78	TB: p. 65 WB: p. 62	Compare fractions with different numerators and denominators.
5	Comparing Fractions — Part 2	p. 80	TB: p. 69 WB: p. 65	Compare fractions to $\frac{1}{2}$ and 1.
6	Practice A	p. 83	TB: p. 73 WB: p. 68	Practice concepts from the chapter.
7	Adding and Subtracting Fractions — Part 1	p. 85	TB: p. 75 WB: p. 71	Add and subtract fractions with the same denominator.
8	Adding and Subtracting Fractions — Part 2	p. 88	TB: p. 80 WB: p. 74	Add and subtract fractions with the same denominator, giving an answer in simplest form.
9	Practice B	p. 90	TB: p. 83 WB: p. 77	Practice adding and subtracting fractions with the same denominator. Practice simplifying fractions.
	Workbook Solutions	p. 92		

Chapter 10 Fractions — Part 2

This chapter builds on Chapter 9: Fractions — Part 1 by introducing equivalent fractions. Students will use their understanding of equivalent fractions to simplify and compare fractions. They will then add and subtract fractions with common denominators.

Equivalent Fractions

Fractions are equal if they represent the same number. The same point on a number line can be labeled with the equivalent fractions.

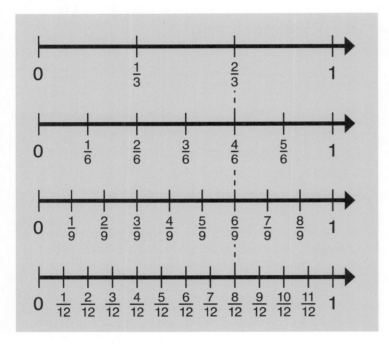

When using bar models to determine whether fractions are equivalent, each bar is the same length, representing the same whole. However, each bar is divided into a different number of equal parts.

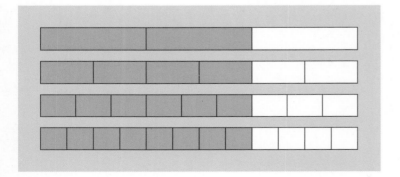

Equivalent fractions can be found by:

- Multiplying the numerator and denominator by the same number.

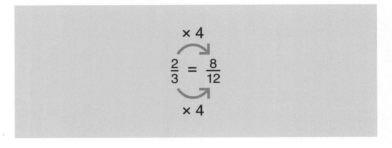

- Dividing the numerator and denominator by the same number.

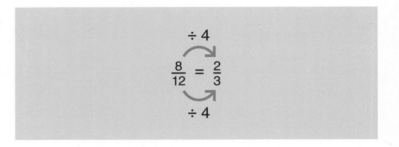

Ensure that students understand the pictorial representations and number line models before introducing the procedure of multiplying the numerator and denominator by the same number.

A common misrepresentation of the process is to show multiplication by a whole number, for example, $\frac{2}{3} \times 4$ does not equal $\frac{8}{12}$.

When we divide the numerator and denominator by the same number, we are simplifying the fraction. If there is no whole number that both the numerator and denominator can be divided by evenly, the fraction is already in simplest form:

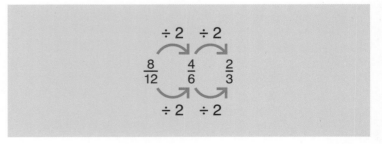

$\frac{2}{3}$ is the simplest form of $\frac{8}{12}$.

Chapter 10 Fractions — Part 2

Notes

Once students have learned to find equivalent fractions, they have the foundation for fraction operations in future grades.

Comparing Fractions

In the previous chapter, students compared fractions with the same numerator or the same denominator. In this chapter, students will compare fractions by finding an equivalent fraction or by comparing to a benchmark fraction. For example:

- Finding an equivalent fraction:

  $\frac{2}{3} > \frac{5}{9}$ because: $\frac{6}{9} > \frac{5}{9}$

- Comparing to 1 whole:

  $\frac{2}{3} < \frac{6}{7}$ because: $\frac{1}{3} > \frac{1}{7}$

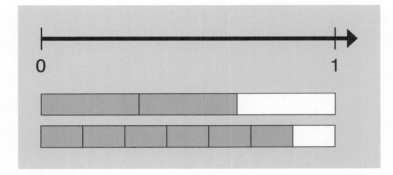

Comparing to the benchmark of 1 can be a challenging concept. Although $\frac{2}{3}$ and $\frac{6}{7}$ are each one unit fraction less than 1, $\frac{1}{3}$ is greater than $\frac{1}{7}$. That means that $\frac{2}{3}$ is further from 1 than $\frac{6}{7}$ is from 1.

- Comparing to $\frac{1}{2}$:

  $\frac{3}{5} > \frac{1}{2}$ because $\frac{1}{2} = \frac{3}{6}$ and $\frac{3}{5} > \frac{3}{6}$

Adding and Subtracting Fractions

Students will add and subtract fractions with common denominators. Students will then learn that some of the answers can be simplified.

$\frac{2}{3} + \frac{1}{3} = \frac{3}{3} = 1$

$\frac{7}{16} - \frac{3}{16} = \frac{4}{16} = \frac{1}{4}$

Addition and subtraction with related denominators such as $\frac{2}{3} + \frac{1}{6}$ will be taught in Dimensions Math 4A. However, students might notice that they can find equivalent fractions from related denominators and add them together.

$\frac{2}{3} + \frac{1}{6}$ can be thought of as: $\frac{4}{6} + \frac{1}{6} = \frac{5}{6}$

$\frac{2}{3} - \frac{1}{6}$ can be thought of as: $\frac{4}{6} - \frac{1}{6} = \frac{3}{6}$

Chapter 10 Fractions — Part 2

Materials

- Blank paper
- Colored pencils
- Dry erase sleeve
- Fraction tiles or fraction circles
- Light colored dry erase marker
- Ruler
- Strips of paper of equal length
- Whiteboards

Blackline Masters

- Equivalent Fractions Puzzle
- Fraction Chart
- Fraction Match Cards
- Multiplication Chart
- Number Cards 1 to 12
- Order Up Cards
- Paint Problem

Activities

Activities included in this chapter are designed to provide practice with finding equivalent fractions. They can be used after students complete the **Do** questions, or any time additional practice is needed.

Chapter Opener

Objective

- Investigate equivalent fractions.

Lesson Materials

- Fraction tiles or fraction circles

Discuss the friends' statements about the pizza.

Ask students, "If the numerators and denominators are all different, how can we compare the amount of pizza each friend ate?"

Provide students with fraction tiles or fraction circles and whiteboards, and have them discuss how they would compare the fractions.

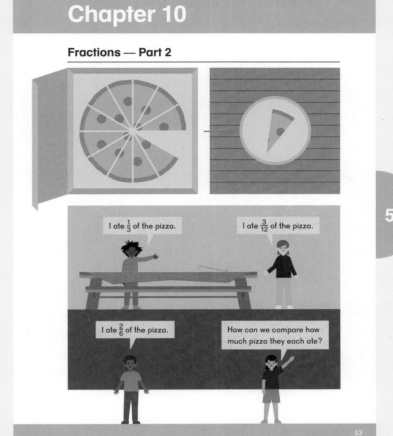

Lesson 1 Equivalent Fractions

Objectives
- Understand the meaning of equivalent fractions.
- Identify equivalent fractions using bar models or number lines.

Lesson Materials
- Strips of paper of equal length, 3 per student
- Fraction tiles
- Fraction Chart (BLM)

Think

Provide each student with three strips of paper and pose the **Think** tasks. Have students fold their paper strips.

Have students try to answer the questions independently. Discuss students' answers to the **Think** questions. Students should see that although the paper strips are folded into different parts, each shaded part is $\frac{1}{2}$ of the whole.

Learn

Dion defines equivalent fractions as fractions that represent the same value. Students may note that the beginning of the word "equivalent" sounds like "equal". Equivalent fractions have the same value, and they name equal points on a number line.

In addition to the fractions equal to $\frac{1}{2}$, students should easily see on the number lines that $\frac{1}{4}$ and $\frac{2}{8}$ as well as $\frac{3}{4}$ and $\frac{6}{8}$ are pairs of equivalent fractions.

Have students save the paper strips for the next lesson.

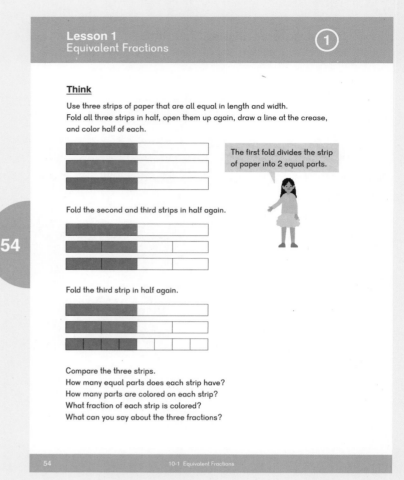

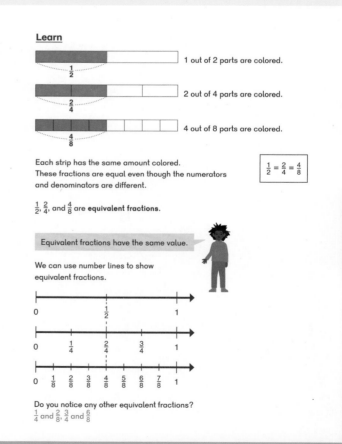

70   Teacher's Guide 3B Chapter 10   © 2017 Singapore Math Inc.

Do

Note that students do not need to multiply or divide numerators and denominators. They can solve the problems using the textbook images.

❹ While students can easily see $\frac{5}{10} = \frac{1}{2}$:

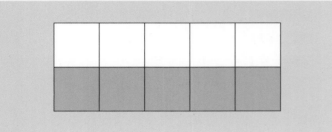

$\frac{4}{10} = \frac{2}{5}$ is harder to visualize this way:

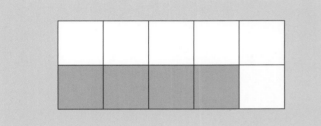

Encourage students to see that fifths are created by the vertical lines here:

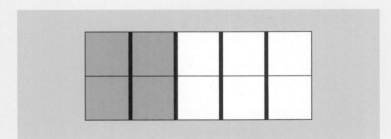

❺ Use Fraction Chart (BLM) or fraction tiles to find equivalent fractions.

Exercise 1 • page 51

Do

❶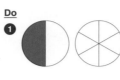

What fraction with a denominator of 6 is equivalent to $\frac{1}{2}$?
$\frac{3}{6}$

❷ $\frac{2}{3}$ of the bar is colored.

(a) $\frac{2}{3} = \frac{4}{6}$

(b) $\frac{2}{3} = \frac{6}{9}$

(c) $\frac{2}{3} = \frac{8}{12}$

❸ (a) What are the missing numbers on the number lines?

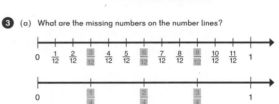

(b) $\frac{3}{12} = \frac{1}{4}$

(c) $\frac{9}{12} = \frac{3}{4}$

(d) $\frac{1}{2} = \frac{6}{12}$

❹ (a) $\frac{5}{10} = \frac{1}{2}$

(b) $\frac{4}{10} = \frac{2}{5}$

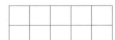

❺ Use the fraction chart to find pairs of equivalent fractions.

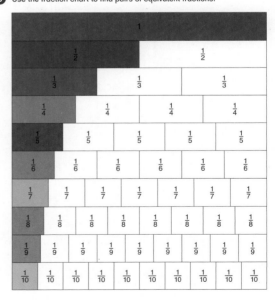

Exercise 1 • page 51

Lesson 2 Finding Equivalent Fractions

Objective

- Find equivalent fractions by multiplying the numerator and denominator by the same number.

Lesson Materials

- Paper strips (same length as in the previous lesson)

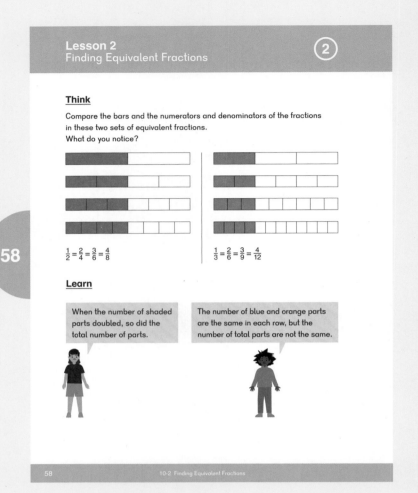

Think

Have students use the paper strips folded into halves, fourths, and eighths from the previous lesson. Provide students with additional strips of paper and have them fold the strips into thirds, sixths, ninths, and twelfths.

Unfold the strips. Lay them out in a similar way to the image in **Think**.

Have students shade their paper strips similarly to the strips in the **Think** problem.

Discuss what students notice about the sets of strips. Then discuss what students notice about the equivalent fractions listed.

Learn

The friends share what they notice about the sets of bars.

Discuss Mei's and Dion's observations. Students should see that as the bars are divided into more parts, there is a corresponding increase in the number of parts that are colored or shaded.

For example:

- When the strip is folded into thirds, 1 of the thirds is shaded.
- When the strip is folded into sixths, 2 of the sixths are shaded.

Ask students if it makes sense that a paper folded into 6 equal parts will have twice the number of equal parts as a paper folded into thirds.

Ask students, "What does Emma mean when she says the size of the shaded part does not change?" Ensure students understand that the length of the shaded part is the same. As a result, the same fraction of the whole is shaded regardless of the number of folds.

Focus the students' attention on Alex's multiplication example and discuss the notation.

We have doubled the total number of parts and the shaded number of parts.

$$\frac{1}{2} \xrightarrow{\times 2} \frac{2}{4} \xrightarrow{\times 2} \frac{4}{8}$$

Ask students if the same method will work for $\frac{1}{3}$.

$$\frac{1}{3} \xrightarrow{\times 2} \frac{2}{6} \xrightarrow{\times 2} \frac{4}{12}$$

Ensure that students understand that finding equivalent fractions is not adding fractions.

A common student misconception is to add both the numerator and denominator together:

$$\frac{1+1}{2+2} = \frac{2}{4}$$

Students should understand that one-half + one-half is two halves: $\frac{2}{2}$, which equals 1.

This understanding becomes even more critical when students begin to add fractions with unlike denominators.

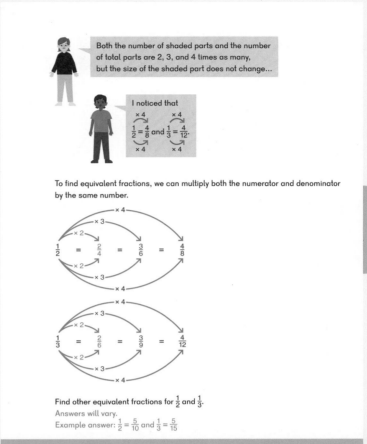

Do

③ (a) As an example, students could first think, "What number can 4 be multiplied by to get a product of 12?" (3)

1 × 3 is 3, so the missing numerator is 3.

(d) Students could first think, "What number can 2 be multiplied by to get a product of 6?" (3)

6 × 3 = 18, so the missing denominator is 18.

Activity

▲ **Investigate Equivalent Fractions on the Multiplication Chart**

Materials: Multiplication Chart (BLM) in a dry erase sleeve, light colored dry erase marker or ruler

A multiplication chart is a good way to further investigate equivalent fractions.

Select two consecutive rows on the chart and have students highlight these or underline them with a ruler.

The first highlighted row represents the numerator and the second highlighted row represents the denominator. ($\frac{2}{3}$) If they move one column to the right (the column that corresponds with "× 2") they will find the equivalent fraction to that found in the first column. ($\frac{4}{6}$)

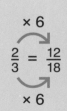

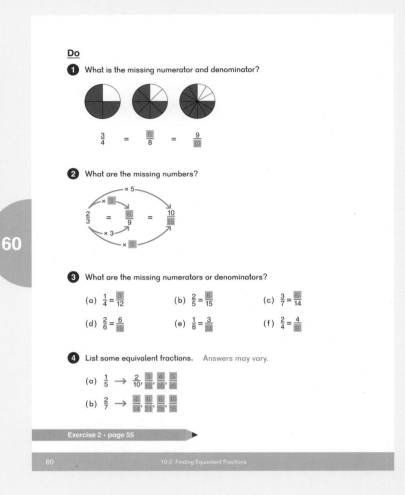

★ Extend the activity by highlighting non-consecutive rows:

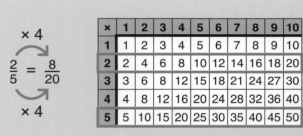

Exercise 2 • page 55

Lesson 3 Simplifying Fractions

Objectives

- Find equivalent fractions by dividing the numerator and denominator by the same number.
- Simplify fractions.

Lesson Materials

- Strips of paper of equal length
- Colored pencils

Think

Provide students with strips of paper of equal length and colored pencils, and pose the **Think** problem. Have students shade in $\frac{8}{12}$ and use that to find and shade equivalent fractions. Have them write the equivalent fractions.

Discuss what students notice about the sets of strips. Then discuss what students notice about the equivalent fractions listed.

Learn

Sofia sees that the numerator and denominator are even and she finds an equivalent fraction by dividing both numbers by 2. She then sees that she can still divide both numbers again by a whole number.

Alex finds an equivalent fraction by dividing by 4:

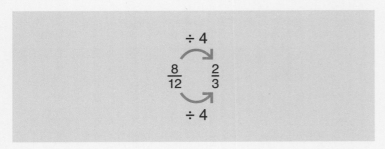

Ask students what is alike and what is different about Sofia's and Alex's methods. They should notice that Alex's method is a shortcut of Sofia's method. Both friends found $\frac{2}{3}$ to be an equivalent fraction of $\frac{8}{12}$.

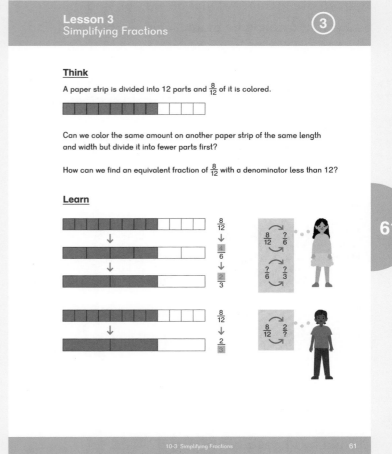

By taking two steps to simplify the fraction, Sofia found $\frac{4}{6}$ as well.

Help students see that just as we can find an equivalent fraction by multiplying, we can also find one by dividing.

To simplify a fraction, we divide the numerator and denominator by the same number. We have decreased the total number of parts and the shaded number of parts.

Dion points out that simpler fractions have fewer parts.

Write some equivalent fractions on the board and ask students if they are in simplest form. If a fraction can be simplified, have students say the fraction in simplest form.

Discuss Mei and Alex's comments.

Do

3 (a) Students should think, "What number can 10 be divided by to get a quotient of 5?" (2)

8 ÷ 2 = 4, so the missing numerator is 4.

(d) Students should first think, "What number can 6 be divided by to get a quotient of 3?" (2)

12 ÷ 2 = 6, so the missing denominator is 6.

4. We cannot show that $\frac{8}{12}$ is equivalent to $\frac{6}{9}$ (without the picture) by dividing both the numerator and denominator of $\frac{8}{12}$ by a whole number. But if both fractions have an equivalent fraction that is the same, $\frac{8}{12}$ and $\frac{6}{9}$ must be the equivalent.

We can find that $\frac{8}{12} = \frac{2}{3}$ and $\frac{6}{9} = \frac{2}{3}$.
Since both fractions are equivalent to $\frac{2}{3}$, we can see that $\frac{8}{12}$ is equivalent to $\frac{6}{9}$.

6. Students may note that if a fraction is equivalent to $\frac{1}{2}$, the denominator is twice the numerator. For example,

$\frac{5}{10}$ 10 = 2 × 5

$\frac{10}{20}$ 20 = 2 × 10

$\frac{50}{100}$ 100 = 2 × 50

Activities

▲ Equivalent Fraction Match

Materials: Fraction Match Cards (BLM)

Lay cards in a faceup array. Have students match two Fraction Match Cards (BLM) that have the same value.

▲ Equivalent Fraction Memory

Materials: Fraction Match Cards (BLM)

Play using the same rules as **Match**, but set the cards out facedown in an array.

Exercise 3 • page 58

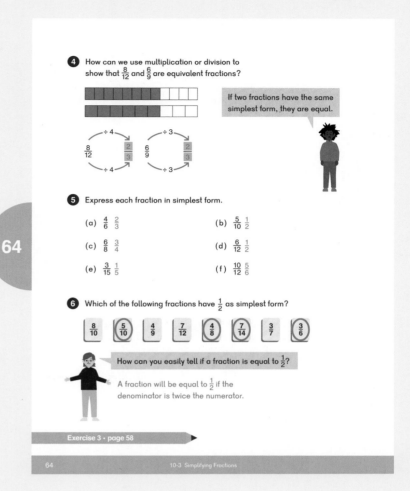

Lesson 4 Comparing Fractions — Part 1

Objective

- Compare fractions with different numerators and denominators.

Think

Discuss the **Think** problem and Sofia's comment. Ask students to recall how they compared fractions with the same numerator or the same denominator.

Provide students with time to find solutions to the **Think** questions.

Discuss students' ideas for comparing the fractions. They will probably suggest finding equivalent fractions.

It is possible that students will try to find a common denominator for fourths, sevenths, and eighths. If so, suggest that they find a way to compare two of the three fractions at a time.

Learn

We can compare fractions that do not have the same numerator or denominator by using equivalent fractions.

Dion sees that the denominator of 4 can be multiplied by 2 to get 8. He finds an equivalent fraction ($\frac{6}{8}$) so that he can compare fractions with common denominators. He compares $\frac{6}{8}$ to $\frac{5}{8}$ and sees that Mei's ribbon is longer than Emma's ribbon.

Emma sees that the numerator of 3 can be multiplied by 2 to get 6 and finds the equivalent fraction of $\frac{6}{8}$. She compares $\frac{6}{8}$ to $\frac{6}{7}$ and sees that $\frac{6}{7}$ is greater than $\frac{6}{8}$, so Dion's ribbon is longer than Mei's ribbon.

Since Mei's ribbon is longer than Emma's, Dion's ribbon must also be longer than Emma's ribbon.

The image at the bottom of Textbook page 66 shows the length of the ribbons.

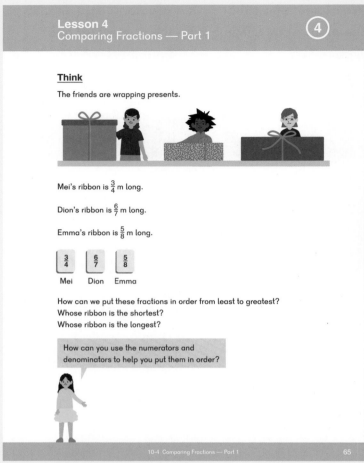

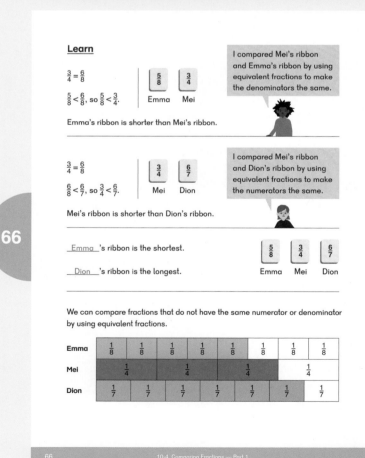

Do

❶–❷ and ❹ Alex, Mei, and Emma think of equivalent fractions to compare.

❶–❷ Alex and Mei compare common denominators.

❹ Emma uses common numerators to compare the fractions.

❺ Sample solutions:

(b) $\frac{3}{5} = \frac{6}{10}$; $\frac{6}{7}$ is greater than $\frac{6}{10}$.

(c) $\frac{2}{7} = \frac{6}{21}$; $\frac{6}{9}$ is greater than $\frac{6}{21}$.

(d) $\frac{9}{8}$ is greater than 1. $\frac{3}{7}$ is less than 1. $\frac{3}{7}$ is less than $\frac{9}{8}$.

(e) $\frac{5}{8} = \frac{10}{16}$; $\frac{10}{11}$ is greater than $\frac{10}{16}$.

❻ This problem suggests what to do when neither a numerator nor denominator are a common multiple of the other. Students can list equivalent fractions systematically until they get to ones with the same numerator or denominator.

❼ Students will need to find an equivalent fraction for both fractions to compare the two fractions. Have students share how they solved these problems.

Exercise 4 • page 62

Lesson 5 Comparing Fractions — Part 2

Objective

- Compare fractions to $\frac{1}{2}$ and 1.

Think

Discuss the **Think** problem and Dion and Emma's questions.

If needed, suggest students draw a picture or use a number line.

Discuss student strategies for comparing the fractions.

Learn

Dion shows that one way to find out that $\frac{2}{5}$ is less than $\frac{1}{2}$ is by finding an equivalent fraction with a common numerator.

Students could also use common denominators to think $\frac{4}{10} < \frac{5}{10}$ so $\frac{2}{5} < \frac{1}{2}$.

Sofia skateboarded less than $\frac{1}{2}$ a mile.

Emma compares the distances skateboarded by Alex and Mei to 1. Both $\frac{5}{6}$ and $\frac{8}{9}$ are more than $\frac{1}{2}$.

The difference between $\frac{5}{6}$ and 1 is more than the difference between $\frac{8}{9}$ and 1.

Mei skateboarded farther than Alex.

The image and number line at the bottom of Textbook page 70 show how far each friend skateboarded.

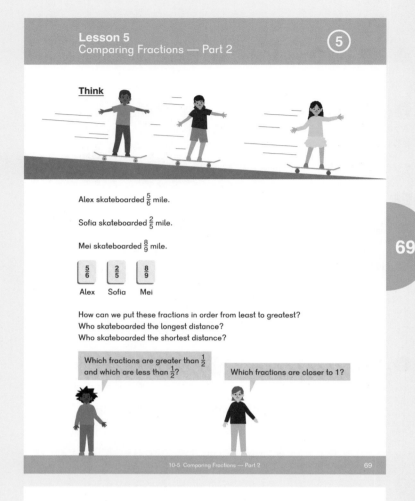

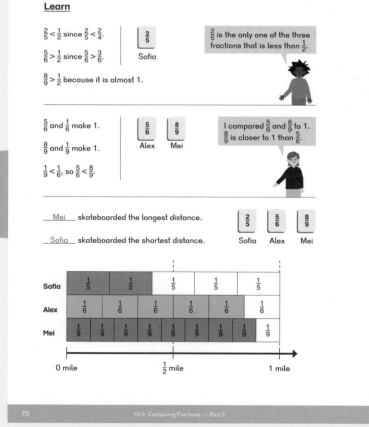

Do

❶—❺ In these problems, students compare fractions to $\frac{1}{2}$.

❶—❸ are scaffolded to compare the fractions $\frac{5}{8}$ and $\frac{4}{9}$ first to $\frac{1}{2}$, then to each other.

❺ Possible solutions:

(a) $\frac{1}{4} < \frac{2}{4}, \frac{5}{7} > \frac{5}{10}$

(b) $\frac{2}{3} > \frac{2}{4}, \frac{1}{5} < \frac{1}{2}$ (or $\frac{2}{10} < \frac{5}{10}$)

(c) $\frac{1}{3} < \frac{1}{2}, \frac{5}{8} > \frac{4}{8}$

Have students share how they solved the problems.

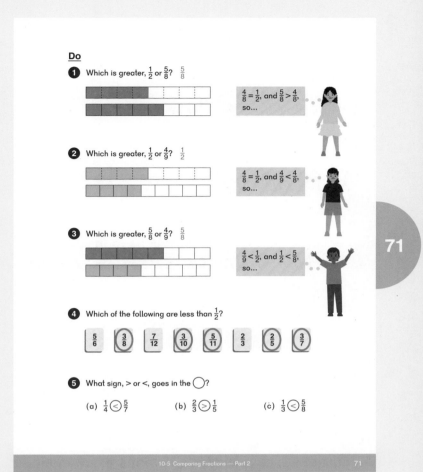

6 – 7 In these problems, students compare fractions to 1.

7 Sample solutions:

(a) $\frac{9}{10}$ is closer to 1 than $\frac{5}{6}$ because $\frac{1}{10} < \frac{1}{6}$.

(b) $\frac{7}{8}$ is closer to 1 than $\frac{2}{3}$ because $\frac{1}{8} < \frac{1}{3}$.

(c) $\frac{3}{4}$ is further from 1 than $\frac{5}{6}$ because $\frac{1}{4} > \frac{1}{6}$.

(d) $\frac{11}{12}$ is closer to 1 than $\frac{8}{9}$ because $\frac{1}{12} > \frac{1}{6}$.

8 Students may compare the fractions to the benchmarks of $\frac{1}{2}$, as Dion does. They may also find equivalent fractions, similar to what Emma does.

Activity

▲ How Many?

Materials: Number Cards (BLM) 1 to 9

Using Number Cards (BLM) 1 to 9 as numerators or denominators, have students determine how many fractions they can make that are less than one-half.

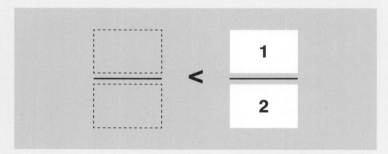

★ Have students find fractions that are between $\frac{1}{4}$ and $\frac{1}{2}$.

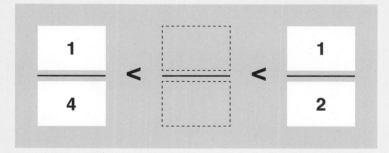

Exercise 5 • page 65

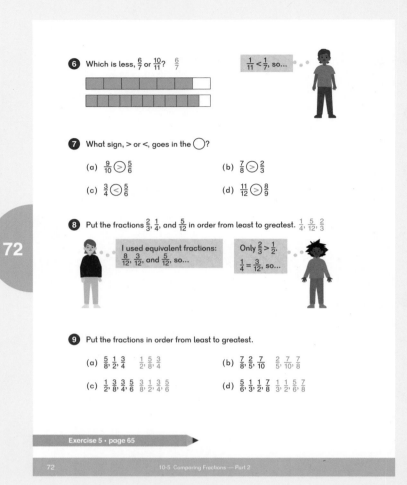

Lesson 6 Practice A

Objective

- Practice concepts from the chapter.

After students complete the **Practice** in the textbook, have them continue to practice ordering and comparing fractions with activities from the chapter.

Activities

▲ Number Line

Materials: Order Up Cards (BLM), string

Repeat the activity from Chapter 9, Lesson 5 on Teacher's Guide page 58 with included equivalent fraction cards.

Tie up the string or have students hold it up at either end. Begin by handing out the 0 and 1 cards to students and have them place the cards on the string number line.

Hand out cards one at a time and have students adjust the other cards in relation to the new cards, place the new cards into their appropriate places:

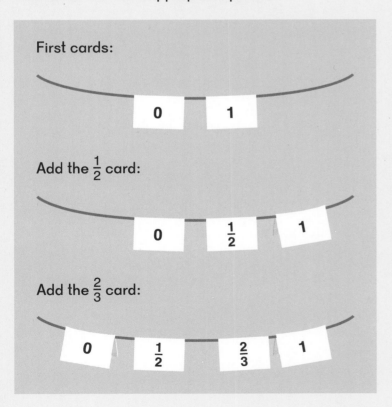

Note that not all cards need to be used for the activity.

① Find the missing numerators.
  (a) $\frac{1}{5} = \frac{2}{10}$ (b) $\frac{6}{9} = \frac{2}{3}$ (c) $\frac{1}{2} = \frac{2}{4} = \frac{3}{6}$
  (d) $\frac{9}{15} = \frac{3}{5}$ (e) $\frac{3}{4} = \frac{6}{8}$ (f) $\frac{1}{3} = \frac{2}{6} = \frac{3}{9}$

② Find the missing denominators.
  (a) $\frac{3}{5} = \frac{6}{10}$ (b) $\frac{2}{3} = \frac{10}{15}$ (c) $\frac{1}{2} = \frac{4}{8} = \frac{6}{12}$
  (d) $\frac{3}{4} = \frac{9}{12}$ (e) $\frac{2}{14} = \frac{1}{7}$ (f) $\frac{2}{3} = \frac{4}{6} = \frac{6}{9}$

③ (a) Write two fractions that are equivalent to $\frac{4}{6}$.
       Answers will vary. Example answer: $\frac{2}{3}, \frac{8}{12}$
    (b) Write two fractions that are equivalent to $\frac{1}{2}$.
       Answers will vary. Example answer: $\frac{2}{4}, \frac{3}{6}$
    (c) Write two fractions that are equivalent to 1.
       Answers will vary. Example answer: $\frac{3}{3}, \frac{5}{5}$

④ What number is indicated by each letter on the number line? Give your answers in simplest form.

A: $\frac{1}{8}$ B: $\frac{3}{16}$ C: $\frac{5}{16}$ D: $\frac{1}{2}$ E: $\frac{9}{16}$ F: $\frac{3}{4}$ G: $\frac{7}{8}$

Have students explain their reasoning when placing their cards on the number line. Equivalent fractions can be stacked on top of each other.

If students struggle, have them discuss where each new card should be placed.

▲ Simplest

Materials: Number Cards (BLM) 1 to 12, several sets

Begin with all of the cards facedown in a pile. In each round, players take turns drawing 8 cards. Players then make as many fractions in simplest form as possible with their sets of cards.

For example, with the following cards:

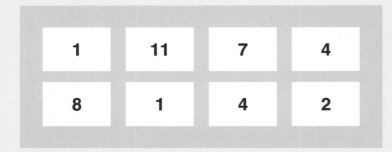

A player could make:

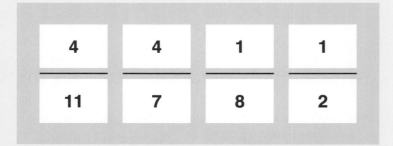

Each fraction in simplest form is worth 1 point. In the example above, the player scores all 4 points.

Players record their points, return the cards to the pile and play another round.

The player with the most points after 5 rounds is the winner.

Extend the activity by having students order their fractions from least to greatest.

Exercise 6 • page 68

Lesson 7 Adding and Subtracting Fractions — Part 1

Objective

- Add and subtract fractions with the same denominator.

Lesson Materials

- Fraction tiles or Paint Problem (BLM)

Think

Provide students with fraction tiles or Paint Problem (BLM) and pose the **Think** problem. Have students show or write their solutions to the questions.

Discuss student solutions.

Learn

Show students how they can use fraction tiles to add $\frac{5}{8}$ and $\frac{2}{8}$.

Emma asks how many one-eighths are in $\frac{5}{8}$ and $\frac{2}{8}$ altogether.

Alex counts on $\frac{2}{8}$ from $\frac{5}{8}$ on a number line.

Mei uses both strategies to answer question (b).

Alex and Mei are using the number word "eighths." Discuss different units with students. Ones, tens, hundreds and thousands can be units. Bananas can be units. Here, "eighths" are the unit. Just as 5 bananas and 2 bananas is 7 bananas, students should see that we are adding the same fractional unit, in this case, eighths.

Dion points out that because the units are the same size, we can simply add or subtract the numerators.

Have students compare their methods from **Think** with the ones in the textbook.

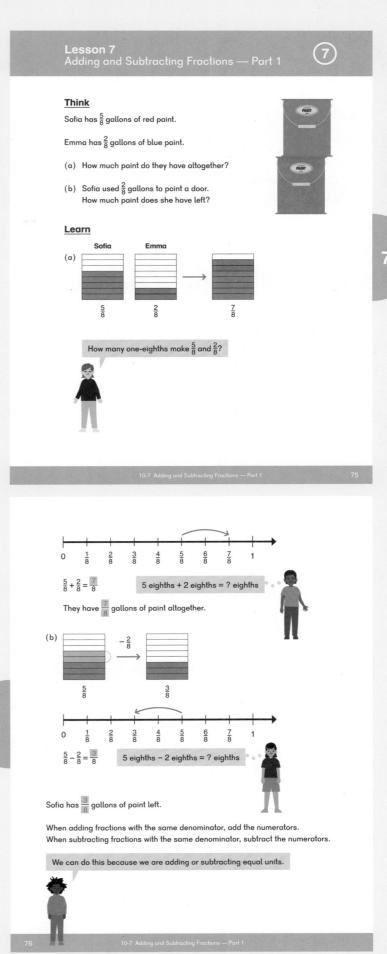

© 2017 Singapore Math Inc.   Teacher's Guide 3B Chapter 10   85

Do

1–**2** Dion and Emma are using the terms "sevenths" and "ninths" to reinforce the idea of adding and subtracting common units.

3 Mei prompts students to think of 1 as 4 fourths.

4 (b) The numerator and denominator are equal, so the answer equals 1.
(e) The calculation results with 0 in the numerator, so the answer equals 0.

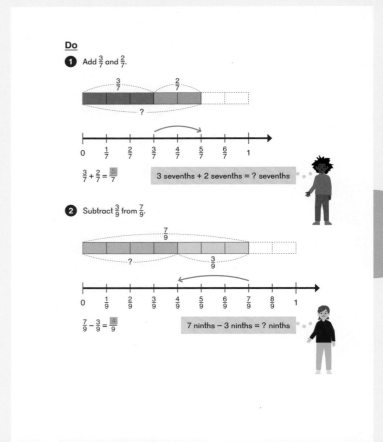

6 (a)–(b) These are missing addend problems. Students who are struggling can think of a number bond, for example:

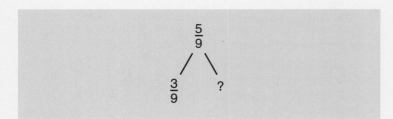

7 Students should think of the lawn as the whole, or 1.

Exercise 7 • page 71

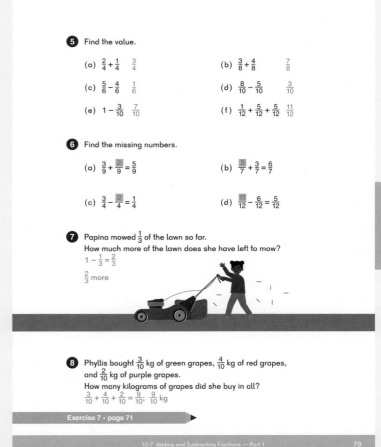

Lesson 8 Adding and Subtracting Fractions — Part 2

Objective

- Add and subtract fractions with the same denominator, giving an answer in simplest form.

Lesson Materials

- Fraction tiles or Paint Problem (BLM)

Think

Provide students with fraction tiles or Paint Problem (BLM) and pose the **Think** problem. Have students show or write their solutions to the questions. If needed, remind students that the question asks for an answer in simplest form.

Discuss student solutions.

Learn

Students could also draw a number line similar to the previous lesson to show the subtraction:

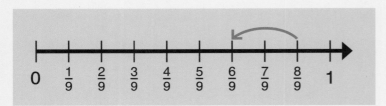

They will then simplify $\frac{6}{9}$ to $\frac{2}{3}$.

Have students compare their methods from **Think** with the ones in the textbook.

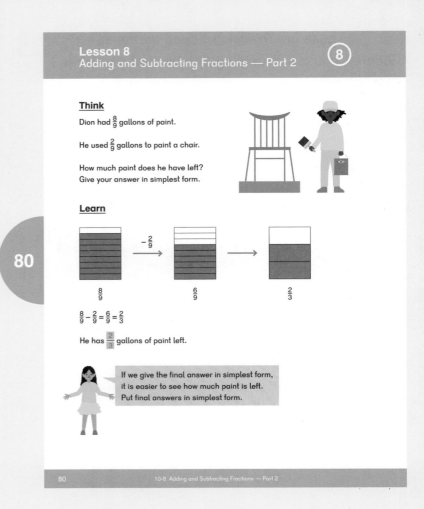

Do

1 – 2 The bar models show how the fractions can first be added and then how the answer can be simplified.

4 The bar models show how the $\frac{2}{6}$ can be subtracted from 1 whole, or $\frac{6}{6}$, then simplified.

Note that in this problem, students might know that $\frac{2}{6} = \frac{1}{3}$ and they can simplify first, then subtract $\frac{1}{3}$ from 1.

Activity

▲ **Word Problems**

Have students write two word problems, one using an addition equation, and another using a subtraction equation from the textbook pages.

Students can trade their word problems with a partner to solve.

Exercise 8 • page 74

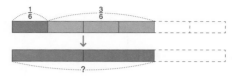
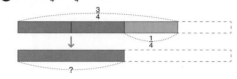
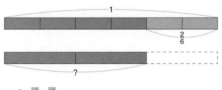

Lesson 9 Practice B

Objectives

- Practice adding and subtracting fractions with the same denominator.
- Practice simplifying fractions.

After students complete the **Practice** in the textbook, have them continue to practice fractions skills with activities from the chapter.

Activity

▲ Snowball Review

Materials: Blank paper

Write large addition and subtraction problems with fractions that have the same denominators on sheets (or half sheets) of paper. Make enough for each student in the class to have 2 or 3.

Have each student crumple up their papers and have a classroom snowball fight for one minute. (No running, safety first!)

At the end of the fight, ask kids to grab a snowball and return to their seats. One at a time, have students unwrap the paper, show the class, and then solve the problem.

When each member of the class has solved one snowball problem, have another "fight" and repeat with snowballs remaining on the floor.

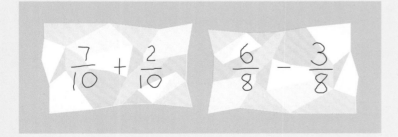

Exercise 9 • page 77

Brain Works

★ Equivalent Fractions Puzzle

Materials: Equivalent Fractions Puzzle (BLM)

To prepare: Print an Equivalent Fractions Puzzle (BLM) on cardstock for each student or pair of students and cut out triangles along the bold lines.

Students match the sides of the triangles with an equivalent fraction.

The puzzle will form a hexagon when complete:

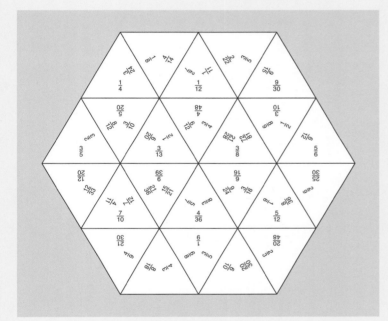

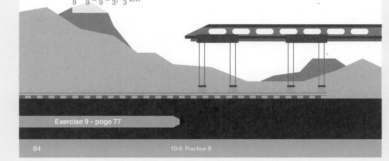

Exercise 1 • pages 51–54

Chapter 10 Fractions — Part 2
Exercise 1

Basics

1. Write equivalent fractions for $\frac{1}{2}$.

   $\frac{1}{2}$ $\boxed{\frac{2}{4}}$ $\boxed{\frac{3}{6}}$ $\boxed{\frac{4}{8}}$ $\boxed{\frac{5}{10}}$ $\boxed{\frac{6}{12}}$

2. Write equivalent fractions for $\frac{3}{4}$.

   $\frac{3}{4}$ $\boxed{\frac{6}{8}}$ $\boxed{\frac{9}{12}}$

3. Use the number lines to find two equivalent fractions for $\frac{2}{3}$.

   $\frac{2}{3}$ $\boxed{\frac{4}{6}}$ $\boxed{\frac{6}{9}}$

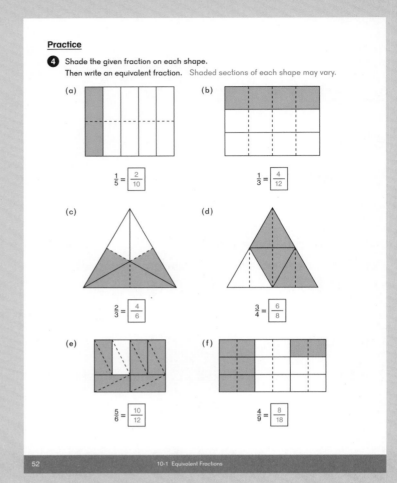

Practice

4. Shade the given fraction on each shape.
   Then write an equivalent fraction. *Shaded sections of each shape may vary.*

   (a) $\frac{1}{5} = \boxed{\frac{2}{10}}$ (b) $\frac{1}{3} = \boxed{\frac{4}{12}}$

   (c) $\frac{2}{3} = \boxed{\frac{4}{6}}$ (d) $\frac{3}{4} = \boxed{\frac{6}{8}}$

   (e) $\frac{5}{6} = \boxed{\frac{10}{12}}$ (f) $\frac{4}{9} = \boxed{\frac{8}{18}}$

5. Label all pairs of equivalent fractions between 0 and 1 shown by the tick marks on each number line.
   One has been done for you.

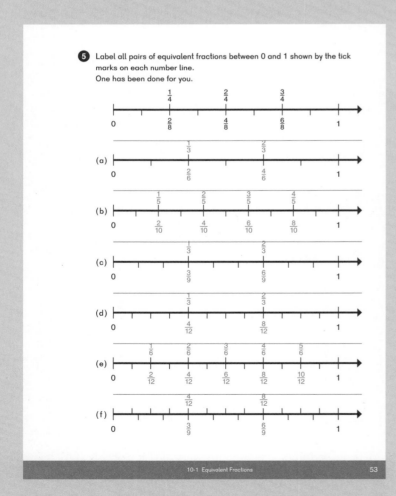

Challenge

6. Shade the given fraction on each figure.
   Then write an equivalent fraction based on the figures.
   Shaded sections of each shape may vary.

   (a) $\frac{3}{8} = \boxed{\frac{6}{16}}$ (b) $\frac{1}{3} = \boxed{\frac{2}{6}}$

   (c) $\frac{1}{2} = \boxed{\frac{5}{10}}$ (d) $\frac{2}{3} = \boxed{\frac{4}{6}}$

   (e) $\frac{3}{4} = \boxed{\frac{6}{8}}$ (f) $\frac{3}{5} = \boxed{\frac{6}{10}}$

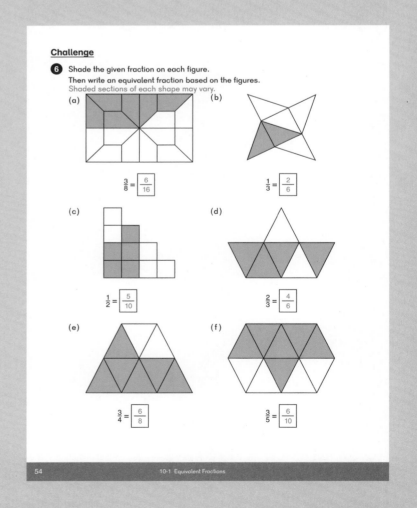

Exercise 2 • pages 55–57

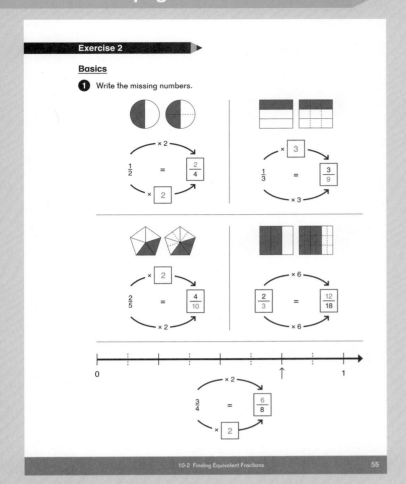

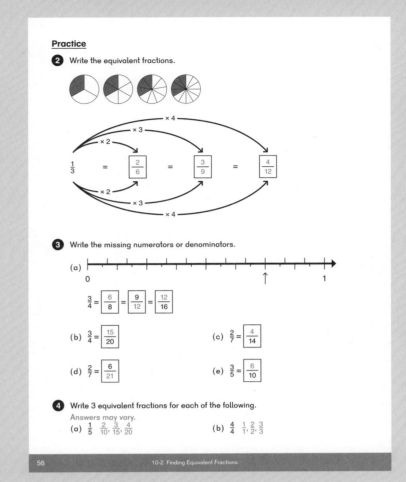

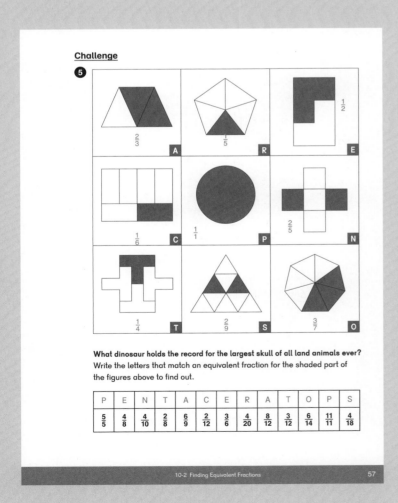

Teacher's Guide 3B Chapter 10

Exercise 3 • pages 58–61

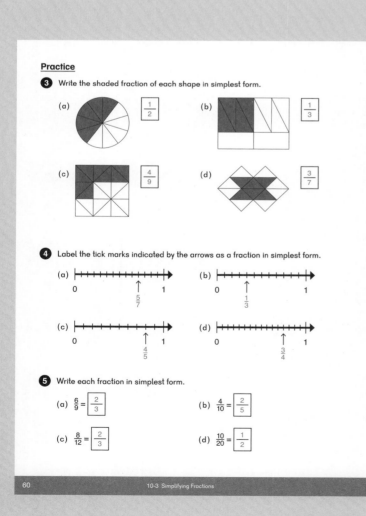
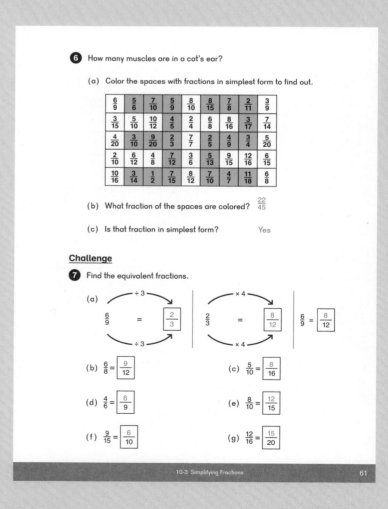

Exercise 4 • pages 62–64

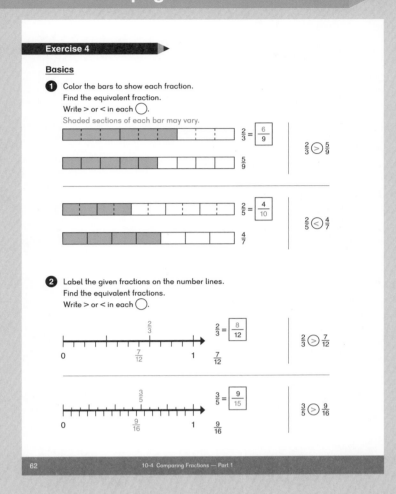

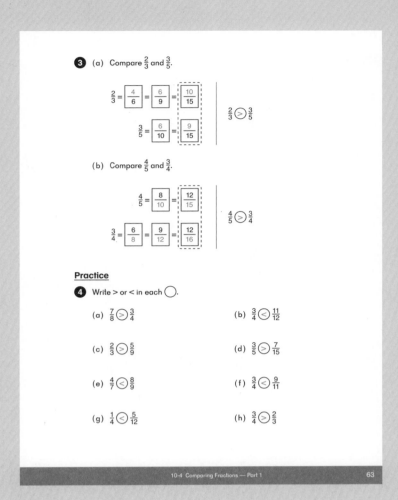

Exercise 5 • pages 65–67

Exercise 5

Basics

1 Color the bars to show each fraction and compare to $\frac{1}{2}$ or 1.
Write >, <, or = in each ◯.

(a) $\frac{5}{8}$, $\frac{4}{9}$

$\frac{1}{2} = \frac{4}{8}$ | $\frac{5}{8} > \frac{4}{8}$, so $\frac{5}{8} > \frac{1}{2}$. | $\frac{5}{8} > \frac{4}{9}$

$\frac{4}{9} < \frac{4}{8}$, so $\frac{4}{9} < \frac{1}{2}$.

(b) $\frac{3}{7}$, $\frac{2}{3}$

$\frac{1}{2} = \frac{7}{14}$ | $\frac{3}{7} = \frac{6}{14}$, and $\frac{6}{14} < \frac{7}{14}$, so $\frac{3}{7} < \frac{1}{2}$. | $\frac{3}{7} < \frac{2}{3}$

$\frac{2}{4} < \frac{2}{3}$ | $\frac{2}{3} > \frac{2}{4}$, so $\frac{2}{3} > \frac{1}{2}$.

(c) $\frac{6}{7}$, $\frac{5}{6}$

$\frac{6}{7}$ and $\frac{1}{7}$ make 1. | $\frac{1}{7} < \frac{1}{6}$, so $\frac{6}{7} > \frac{5}{6}$.

$\frac{5}{6}$ and $\frac{1}{6}$ make 1.

Practice

2 Circle all of the fractions that are less than $\frac{1}{2}$.

$\frac{2}{6}$ $\frac{4}{6}$ $\frac{3}{8}$ $\frac{5}{8}$ $\frac{4}{10}$ $\frac{6}{10}$ $\frac{5}{12}$ $\frac{7}{12}$ $\frac{7}{16}$ $\frac{9}{16}$

3 Circle all of the fractions that are greater than $\frac{1}{2}$.

$\frac{3}{5}$ $\frac{3}{7}$ $\frac{4}{7}$ $\frac{4}{9}$ $\frac{5}{9}$ $\frac{5}{11}$ $\frac{6}{11}$ $\frac{6}{13}$ $\frac{7}{13}$ $\frac{7}{15}$

4 Write > or < in each ◯.

(a) $\frac{3}{8} < \frac{5}{9}$ (b) $\frac{5}{8} > \frac{7}{12}$

(c) $\frac{2}{3} > \frac{5}{8}$ (d) $\frac{3}{7} < \frac{8}{15}$

(e) $\frac{5}{7} > \frac{3}{8}$ (f) $\frac{3}{4} > \frac{5}{11}$

5 Write > or < in each ◯.

(a) $\frac{7}{8} > \frac{6}{7}$ (b) $\frac{4}{5} < \frac{11}{12}$

(c) $\frac{2}{3} < \frac{4}{5}$ (d) $\frac{6}{7} < \frac{9}{10}$

(e) $\frac{6}{7} < \frac{8}{9}$ (f) $\frac{10}{11} > \frac{8}{9}$

6 Write the fractions in order from least to greatest.

(a) $\frac{5}{6}, \frac{6}{7}, \frac{2}{5}$ $\frac{2}{5}, \frac{5}{6}, \frac{6}{7}$

(b) $\frac{7}{12}, \frac{3}{4}, \frac{3}{7}$ $\frac{3}{7}, \frac{7}{12}, \frac{3}{4}$

(c) $\frac{5}{9}, \frac{2}{5}, \frac{13}{12}$ $\frac{2}{5}, \frac{5}{9}, \frac{13}{12}$

(d) $\frac{9}{11}, \frac{3}{7}, \frac{3}{10}, \frac{8}{8}$ $\frac{3}{10}, \frac{3}{7}, \frac{9}{11}, \frac{8}{8}$

(e) $\frac{8}{9}, \frac{1}{2}, \frac{3}{7}, \frac{7}{8}$ $\frac{3}{7}, \frac{1}{2}, \frac{7}{8}, \frac{8}{9}$

Challenge

7 Use the given numbers to fill in the missing numerators or denominators so that the fractions are in order from least to greatest.
Each fraction should be less than 1.

(a) 7, 8, 9 $\frac{4}{7} < \frac{7}{8} < \frac{8}{9}$

(b) 2, 3, 4 $\frac{2}{7} < \frac{4}{9} < \frac{3}{4}$

(c) 6, 7, 8 $\frac{3}{8} < \frac{7}{9} < \frac{6}{7}$

Exercise 6 • pages 68–70

Exercise 6

Check

1. Which figure does not correctly show $\frac{3}{4}$ shaded? **B**
   Explain why the other figures are $\frac{3}{4}$ shaded.

   A
   $\frac{3}{4}$ of this figure is shaded.

   B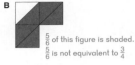
   $\frac{5}{6}$ of this figure is shaded.
   $\frac{5}{6}$ is not equivalent to $\frac{3}{4}$

   C
   $\frac{9}{12}$ is an equivalent fraction of $\frac{3}{4}$.

   D
   $\frac{9}{12}$ is an equivalent fraction of $\frac{3}{4}$.

2. Label all pairs of equivalent fractions between 0 and 1 shown by the tick marks on each number line.

   (a)

   (b)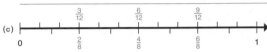

   (c)

3. Circle all fractions greater than 1.

   $\frac{7}{9}$ ⓕ$\frac{6}{4}$ $\frac{3}{3}$ ⓕ$\frac{8}{5}$ ⓕ$\frac{10}{9}$ $\frac{7}{10}$ ⓕ$\frac{20}{16}$ $\frac{15}{20}$

4. Write 4 equivalent fractions for each of the following.
   Answers may vary.
   (a) $\frac{2}{3}$ $\frac{4}{6}, \frac{6}{9}, \frac{8}{12}, \frac{10}{15}$
   (b) $\frac{3}{5}$ $\frac{6}{10}, \frac{9}{15}, \frac{12}{20}, \frac{15}{25}$

5. Fill in the numerators or denominators to write equivalent fractions for $\frac{1}{2}$.
   (a) $\frac{6}{12}$ (b) $\frac{3}{6}$
   (c) $\frac{5}{10}$ (d) $\frac{8}{16}$

6. Write each fraction in its simplest form.
   (a) $\frac{6}{8} = \frac{3}{4}$ (b) $\frac{8}{10} = \frac{4}{5}$
   (c) $\frac{9}{12} = \frac{3}{4}$ (d) $\frac{10}{16} = \frac{5}{8}$

7. $\frac{3}{10}$ of a bag of flour was used to bake bread and $\frac{1}{4}$ of the bag was used to make scones.
   In which baked good was more flour used?
   $\frac{3}{10} = \frac{6}{20}$
   $\frac{1}{4} = \frac{5}{20}$
   More flour was used for the bread.

8. Label the tick marks on the number lines with the given fractions.

   (a) $\frac{7}{8}, \frac{2}{3}, \frac{6}{7}, \frac{2}{5}$

   0 — $\frac{2}{5}$ — $\frac{1}{2}$ — $\frac{2}{3}$ — $\frac{6}{7}$ $\frac{7}{8}$ — 1

   (b) $\frac{3}{7}, \frac{8}{9}, \frac{3}{10}, \frac{5}{9}$

   0 — $\frac{3}{10}$ — $\frac{3}{7}$ $\frac{1}{2}$ $\frac{5}{9}$ — $\frac{8}{9}$ — 1

9. Write the missing denominators.
   (a) $\frac{1}{4} < \frac{1}{3} < \frac{1}{2}$
   (b) $\frac{2}{9} < \frac{1}{4} < \frac{2}{7}$

Challenge

10. Write the missing denominator.
    $\frac{2}{7} < \frac{1}{3} < \frac{3}{8}$

11. The amount of water in a tank doubles each day.
    If the tank becomes full on June 5, on what day was it half full?
    June 4

Exercise 7 • pages 71–73

Exercise 7

Basics

1. Color each of the two fractions a different color on the figures. Then add the fractions.

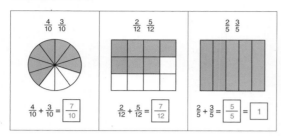

$\frac{4}{10} + \frac{3}{10} = \boxed{\frac{7}{10}}$ $\frac{2}{12} + \frac{5}{12} = \boxed{\frac{7}{12}}$ $\frac{2}{5} + \frac{3}{5} = \boxed{\frac{5}{5}} = \boxed{1}$

2. Show the addition on each number line, then add the fractions. The first number line has been done for you.

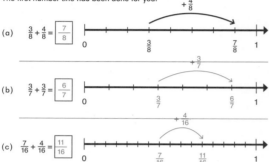

(a) $\frac{3}{8} + \frac{4}{8} = \boxed{\frac{7}{8}}$

(b) $\frac{3}{7} + \frac{3}{7} = \boxed{\frac{6}{7}}$

(c) $\frac{7}{16} + \frac{4}{16} = \boxed{\frac{11}{16}}$

3. Use the bars to help you subtract.

(a) $\frac{8}{9} - \frac{4}{9} = \boxed{\frac{4}{9}}$

(b) $\frac{8}{10} - \frac{5}{10} = \boxed{\frac{3}{10}}$

(c) $1 - \frac{5}{6} = \boxed{\frac{1}{6}}$

4. Show the subtraction on each number line, then subtract the fractions. The first number line has been partly done for you.

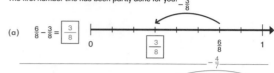

(a) $\frac{6}{8} - \frac{3}{8} = \boxed{\frac{3}{8}}$

(b) $1 - \frac{4}{7} = \boxed{\frac{3}{7}}$

(c) $\frac{11}{12} - \frac{6}{12} = \boxed{\frac{5}{12}}$

(d) $\frac{3}{4} - \frac{3}{4} = \boxed{\frac{0}{4}}$ or 0

Practice

5. Add or subtract.

(a) $\frac{1}{7} + \frac{3}{7} = \boxed{\frac{4}{7}}$ (b) $\frac{3}{6} + \frac{3}{6} = \boxed{\frac{6}{6}}$ or 1

(c) $\frac{8}{9} - \frac{7}{9} = \boxed{\frac{1}{9}}$ (d) $\frac{10}{16} - \frac{5}{16} = \boxed{\frac{5}{16}}$

(e) $1 - \frac{5}{12} = \boxed{\frac{7}{12}}$ (f) $\frac{3}{9} + \frac{1}{9} + \frac{3}{9} = \boxed{\frac{7}{9}}$

(g) $\frac{1}{4} + \frac{1}{4} + \frac{1}{4} + \frac{1}{4} = \boxed{\frac{4}{4}}$ or 1   (h) $\frac{11}{15} - \frac{4}{15} - \frac{3}{15} = \boxed{\frac{4}{15}}$

6. Ivy made some trail mix using $\frac{4}{8}$ cup of nuts, $\frac{2}{8}$ cup of raisins, and $\frac{1}{8}$ cup of chocolate chips.
How many cups of trail mix does she have in all?
$\frac{4}{8} + \frac{2}{8} + \frac{1}{8} = \frac{7}{8}$

She has $\frac{7}{8}$ cups of trail mix in all.

7. Isaac has a board that is 1 meter long.
He sawed off a piece $\frac{4}{10}$ m long and another piece $\frac{3}{10}$ m long.
How long is the remaining piece?
$\frac{10}{10} - \frac{4}{10} - \frac{3}{10} = \frac{3}{10}$

The remaining piece is $\frac{3}{10}$ m.

Exercise 8 • pages 74–76

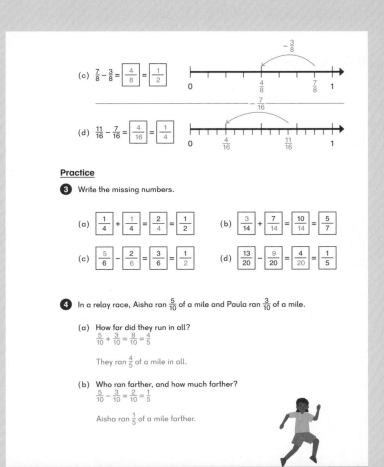

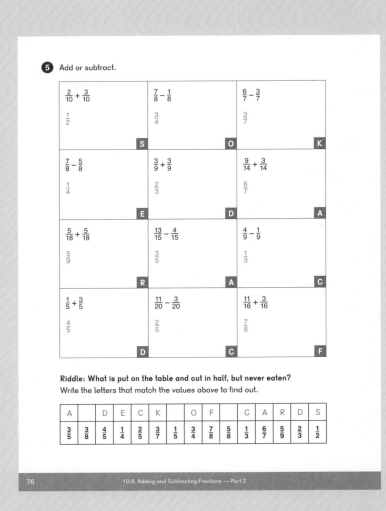

Exercise 9 • pages 77–80

Exercise 9

Check

1 Find the values.

3,468 + 2,781 6,249	8,419 − 672 7,747	4,009 − 2,145 1,864
895 × 3 2,685	156 ÷ 5 31 R 1	665 × 7 4,655
96 ÷ 6 16	634 ÷ 9 70 R 4	888 × 2 1,776
230 + 987 + 158 1,375	1,569 + 3,871 + 78 5,518	45 + 2,409 + 9 + 348 2,811

2 How many more triangles must be shaded to show the given fractions? Complete the equations to show the sum of the shaded part and the part that still needs to be shaded.

(a)

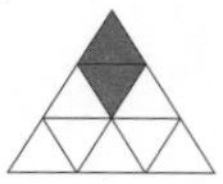

Shade 4 more parts to have $\frac{2}{3}$ shaded.

$\frac{2}{9} + \frac{4}{9} = \frac{2}{3}$

(b)

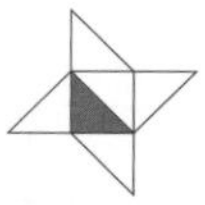

Shade 2 more parts to have $\frac{1}{2}$ shaded.

$\frac{1}{6} + \frac{2}{6} = \frac{1}{2}$

(c)

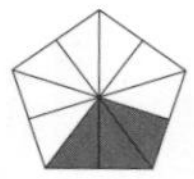

Shade 5 more parts to have $\frac{4}{5}$ shaded.

$\frac{3}{10} + \frac{5}{10} = \frac{4}{5}$

(d)

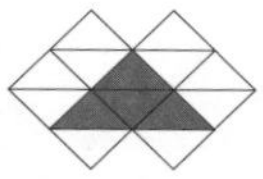

Shade 8 more parts to have $\frac{6}{7}$ shaded.

$\frac{4}{14} + \frac{8}{14} = \frac{6}{7}$

(e)

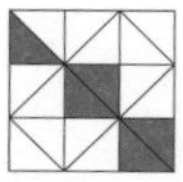

Shade 7 more parts to have $\frac{2}{3}$ shaded.

$\frac{5}{18} + \frac{7}{18} = \frac{2}{3}$

(f)

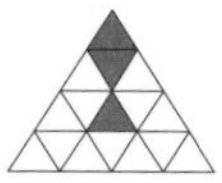

Shade 9 more parts to have $\frac{3}{4}$ shaded.

$\frac{3}{16} + \frac{9}{16} = \frac{3}{4}$

3 Complete the equations for the subtraction shown on each number line.

(a)

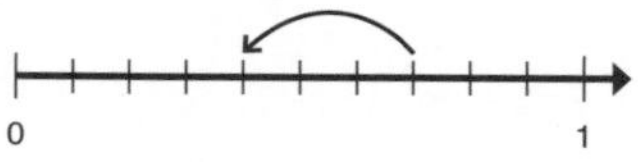

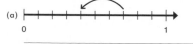

$\frac{7}{10} - \frac{3}{10} = \frac{2}{5}$

(b)

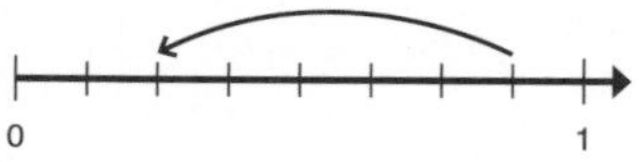

$\frac{7}{8} - \frac{5}{8} = \frac{1}{4}$

(c)

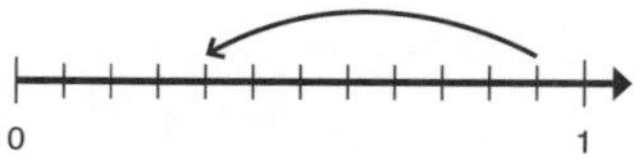

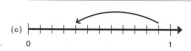

$\frac{11}{12} - \frac{7}{12} = \frac{1}{3}$

4 Write >, <, or = in each ◯.

(a) $\frac{1}{7} + \frac{6}{7} > \frac{1}{9} + \frac{6}{9}$

(b) $\frac{4}{5} - \frac{4}{5} < \frac{7}{8} - \frac{4}{8}$

(c) $\frac{1}{3} + \frac{1}{3} > \frac{9}{12} - \frac{4}{12}$

(d) $\frac{1}{12} + \frac{3}{12} = 1 - \frac{4}{6}$

(e) $\frac{3}{10} + \frac{2}{10} = \frac{13}{16} - \frac{5}{16}$

(f) $\frac{5}{12} + \frac{5}{12} > \frac{1}{4} + \frac{1}{4} + \frac{1}{4}$

(g) $\frac{1}{8} + \frac{3}{8} + \frac{1}{8} > \frac{4}{5} - \frac{3}{5}$

(h) $1 - \frac{1}{3} > \frac{2}{9} + \frac{2}{9}$

Challenge

5 Which is greater, $\frac{2}{5}$ or $\frac{7}{21}$?

$\frac{7}{21} = \frac{1}{3}$, $\frac{2}{5} > \frac{2}{6}$

$\frac{2}{5}$

6 Show how 4 chocolate bars can be shared equally among 5 people.

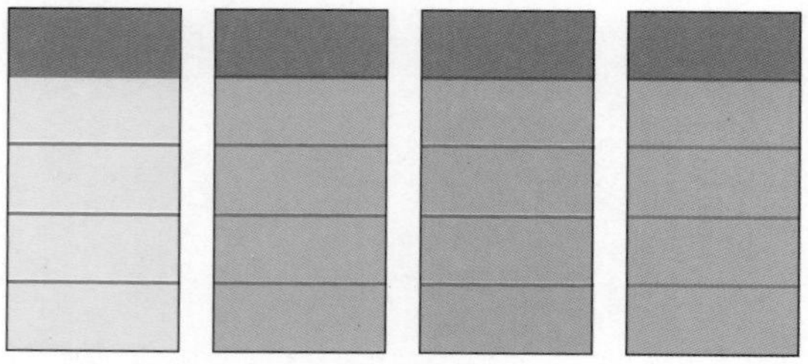

Divide each bar into 5 pieces. That makes 20 total pieces.

20 ÷ 5 = 4

Each person can have 4 pieces, or $\frac{4}{5}$ of a bar.

7 Use straight lines to divide this figure into fourths. Each fourth should be the same shape and size.

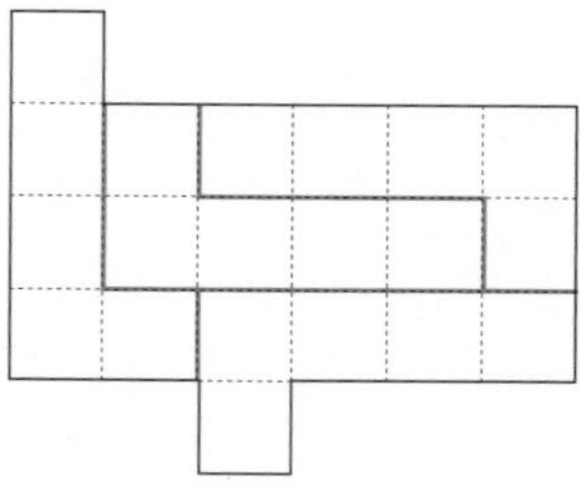

Chapter 11 Measurement — Overview

Suggested number of class periods: 9–10

	Lesson	Page	Resources	Objectives
	Chapter Opener	p. 105	TB: p. 85	Start to understand the usefulness of expressing measurement in compound units.
1	Meters and Centimeters	p. 106	TB: p. 86 WB: p. 81	Express lengths in meters and centimeters.
2	Subtracting from Meters	p. 109	TB: p. 90 WB: p. 85	Subtract centimeters or meters and centimeters from a whole number of meters.
3	Kilometers	p. 111	TB: p. 94 WB: p. 88	Understand and use the kilometer as a unit to measure long distances. Convert between kilometers and meters. Distinguish between travel distance and direct distance.
4	Subtracting from Kilometers	p. 113	TB: p. 98 WB: p. 91	Subtract meters or kilometers and meters from a whole number of kilometers.
5	Liters and Milliliters	p. 115	TB: p. 102 WB: p. 94	Understand capacity and liquid volume. Measure capacity in liters and milliliters.
6	Kilograms and Grams	p. 118	TB: p. 107 WB: p. 97	Understand weight and mass. Measure mass in grams and kilograms.
7	Word Problems	p. 120	TB: p. 111 WB: p. 101	Solve word problems involving measurement.
8	Practice	p. 124	TB: p. 116 WB: p. 105	Practice adding and subtracting compound units of measurement.
	Review 3	p. 126	TB: p. 119 WB: p. 109	Review topics from Chapter 1 through Chapter 11.
	Workbook Solutions	p. 129		

Chapter 11 Measurement

In Dimensions Math 2, some standard units of measurement (centimeter, meter, inch, foot, gram, kilogram, pounds, and liter) were introduced:

- The cat is 15 inches long.
- The rock weighs 10 grams.

In this chapter, compound units are introduced. Students learn to measure in compound units (e.g., 4 m 30 cm) and convert between compound units and a base unit (e.g., 4 m 30 cm = 430 cm).

At each step, students solve both calculation and word problems that build in complexity. These problems allow students to practice and apply the mental math strategies from previous lessons. Since students will only be looking at compound units in the metric system at this level, these lessons reinforce the base-ten place value system.

Length

In Dimensions Math 2, students learned to measure, compare, and estimate lengths using centimeters, meters, inches, and feet. In this chapter, students learn about a new unit of measurement, kilometers, and that:

- 100 centimeters = 1 meter
- 1,000 meters = 1 kilometer

Students will convert between the compound units and a base unit.

Capacity

In Dimensions Math 2, students learned to measure, compare, and estimate capacity using liters. In this chapter, students learn a new unit of measurement, milliliters, and that:

- 1,000 milliliters = 1 liter

Note that the capacity of a container is only an approximation as liquids generally only fill to a fill line.

However, we can still compare the capacities and sizes of different containers.

Later, in Dimensions Math 5, students will learn to measure volume in cubic units, such as cubic centimeters. Volume is the amount of space an object takes up. The capacity of a container is its volume, at least to the fill mark measured in milliliters. When water is poured into a measuring container, such as a beaker, we are measuring the volume of that water in milliliters.

Using water to compare capacity allows students to begin to get an idea of what volume is (the amount of space an object occupies), without being restricted to cubes. A formal definition of volume is not given at this level, since students are not learning about cubic units yet. Instead, the textbook simply says that the amount of liquid in the container is the volume of the liquid, and uses the term "liquid volume."

Weight

In Dimensions Math 2, students learned to measure, compare, and estimate weights using grams, kilograms, and pounds. In this chapter, students learn that:

- 1,000 grams = 1 kilogram

Weight and mass are different concepts, though in normal life we don't worry about this distinction. The metric system measures mass but standard English units measure weight.

Grams and kilograms are measures of mass. Mass is a measure of the amount of matter in that object. It does not depend on gravity. Pounds and ounces are a measure of weight. Weight is the quantity of force that gravity exerts on an object, given its mass. On Earth, a 1-kilogram mass weighs 2.2 pounds.

Students should understand that size does not

Chapter 11 Measurement

determine if something is light or heavy. A pillow may be lighter than a rock even though it may be larger than the rock.

Much of the work in this chapter involves measuring and weighing objects in the classroom. Students need the following tools:

- Platform scale, one per group of 3 to 4
- Weights: 1 g, 10 g, 100 g, 500 g, and 1 kg

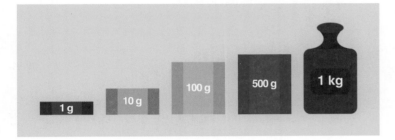

Students should become familiar enough with units of measurement to make some rough estimates or determine whether a measurement is reasonable. They should learn that all measurements are approximations. How close the approximation is to the actual measurement depends on the size of the unit.

Customary units of measurement, such as miles, quarts, and gallons, will be used in word problems. It is assumed that students are familiar with these units of measurements from their daily lives. They will not be measuring or converting compound units and a base unit for these units.

Chapter 11 Measurement — Materials

Materials

- 1,000 mL beakers
- 500 mL beakers
- Centimeter rulers
- Chalk or painter's tape
- Inch rulers
- Index cards
- Items that weigh between 1 kg and 5 kg
- Large pieces of butcher paper
- Measuring tape
- Meter sticks
- Platform scales measuring in metric units
- Small paper cups
- Teaspoons
- Water
- Water bottles ($\frac{1}{2}$ liter)
- Weight sets
- Yardsticks
- Whiteboards

Blackline Masters

- Measurement Match Cards

Activities

Fewer games and activities are included in this chapter, as students will be using measuring tools. The included activities can be used after students complete the **Do** questions, or anytime additional practice is needed.

Chapter Opener

Objective
- Start to understand the usefulness of expressing measurement in compound units.

Discuss the different ways that students are expressing their measurements in meters and centimeters. Review the abbreviations for centimeter (cm) and meter (m).

Use the activity below to extend to a full lesson or continue to Lesson 1: Meters and Centimeters.

Activity

▲ **Measure Me!**

Materials: Large pieces of butcher paper, meter sticks, centimeter rulers, yardsticks, inch rulers

Have students work with partners to trace around and create an outline of their bodies. They can record different measurements in centimeters or meters. "My head is 15 centimeters across. My right arm is 8 centimeters across," etc.

Extend the activity by having students measure in inches and feet with yardsticks and inch rulers.

Chapter 11

Measurement

- I am between 1 m and 2 m tall. I am about $\frac{1}{4}$ m more than 1 m tall, or $\frac{5}{4}$ m.
- I wonder how tall I am in centimeters.
- I know there are 100 cm in 1 m. I could give my height as 1 m and some centimeters.

© 2017 Singapore Math Inc. Teacher's Guide 3B Chapter 11

Lesson 1 Meters and Centimeters

Objective
- Express lengths in meters and centimeters.

Lesson Materials
- Meter sticks

Think

Provide pairs of students with meter sticks and pose the **Think** problem.

Discuss student solutions to the **Think** questions. Students should see that although the heights are expressed in different units, they can be changed to like units to be compared.

Learn

(a) Alex tells us there are 100 centimeters in 1 meter.

The bar model and equations show how we can convert 128 cm into meters and centimeters by splitting it into 100 cm and left over centimeters.

(b) The bar model and equations show how we can convert a measurement in meters and centimeters to centimeters only by using 1 m = 100 cm and adding the centimeters.

Mei provides the definition of compound units.

We can compare Alex's and Emma's heights in compound units.

Alex is 1 m 28 cm tall.
Emma is 1 m 30 cm tall.

We can also compare them in centimeters:

Alex is 128 cm tall.
Emma is 130 cm tall.

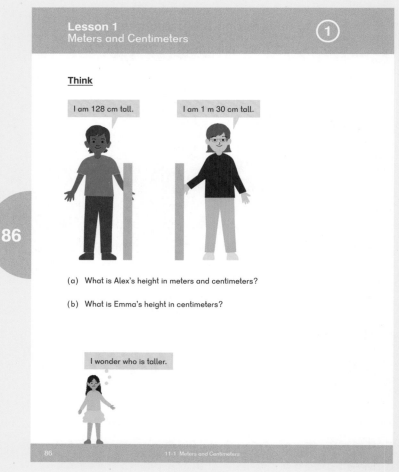

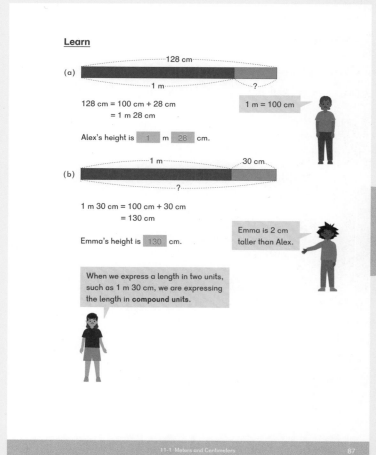

106   Teacher's Guide 3B Chapter 11   © 2017 Singapore Math Inc.

Do

If needed, point out to students that:

1 100 cm = 1 m
2 × 100 cm = 200 cm = 2 m
3 × 100 cm = 300 cm = 3 m
4 × 100 cm = 400 cm = 4 m

2 1 m = 100 cm
2 m = 2 × 100 cm = 200 cm
5 m = 5 × 100 cm = 500 cm

3 Mei splits 165 cm into 100 cm and 65 cm.
She knows 100 cm = 1 m.

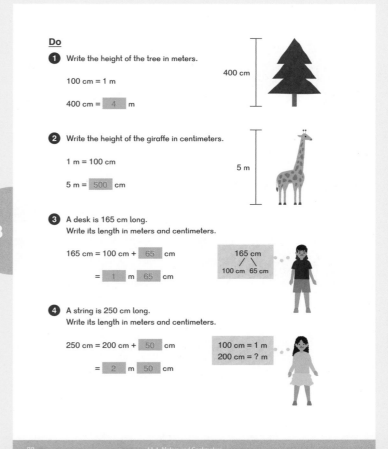

7 (c) Ensure that students understand that:

7 m 5 cm = 700 cm + 5 cm

A common student error is to think of this as 75 cm or as 750 cm.

Activities

▲ Estimate Length

Have students find items around the classroom. Have them estimate whether the items they found are more or less than 1 meter long. Decide which are best measured in centimeters and which are best measured in compound units.

▲ How Far?

Materials: Measuring tape, chalk or painter's tape, index cards

Have students mark off a straight line in meters to measure how far their paper airplanes fly. If outdoors, use chalk. If indoors, use painter's tape in a hallway or classroom.

Teach students how to fold a basic paper airplane. Have them start on the zero line and throw their airplanes. Students can measure the distance their airplanes fly and record it using compound units.

Exercise 1 • page 81

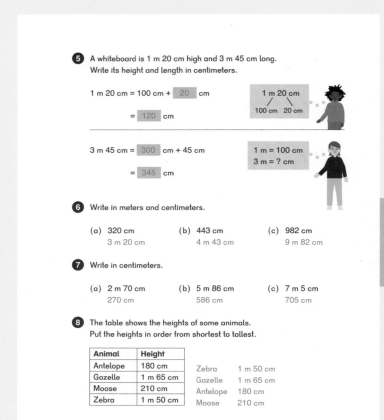

Lesson 2 Subtracting from Meters

Objective

- Subtract centimeters or meters and centimeters from a whole number of meters.

Think

Pose the **Think** problem and have students work independently.

Discuss Alex, Emma, and Mei's thoughts.

Learn

Method 1

Alex splits 3 m into 2 m and 100 cm. He subtracts 75 cm from 100 cm and adds 2 m.

Method 2

Since 1 m is 25 cm more than 75 cm, Emma subtracts 1 m and adds 25 cm.

Method 3

Sofia converts 3 m to 300 cm. She subtracts 75 m from 300 m, and then converts 225 cm to 2 m 25 cm.

Have students compare their methods used in the **Think** with the ones in the textbook.

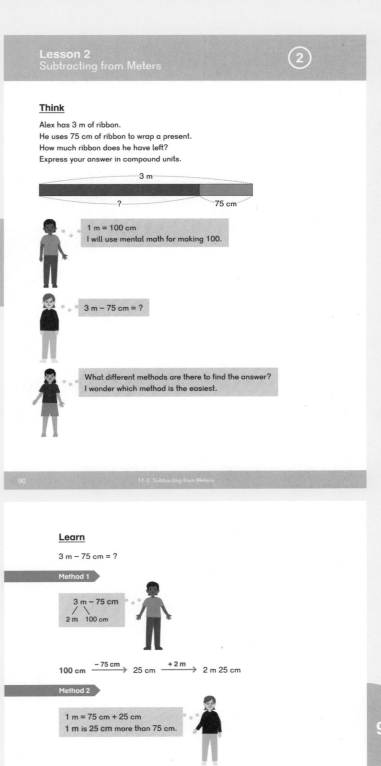

Do

① This problem could be used as an activity by writing these pairs and similar sums to 100 on index cards.

② (b) Emma uses Sofia's method. She knows 3 m is 2 m 100 cm. She subtracts 35 cm from 100 cm and adds 2 m back.

If you can subtract from 1 m, you can subtract from any whole number of meters by simply subtracting from one of the meters.

(c) Mei uses both Sofia and Emma's methods. She subtracts the meters first, then she subtracts the centimeters from the meters.

(d) Dion over-subtracts. Because 3 m 95 cm is very close to 4 m, he subtracts 4 m then adds back 5 cm.

③ — ④ Students can use any strategy to find the answer.

Exercise 2 • page 85

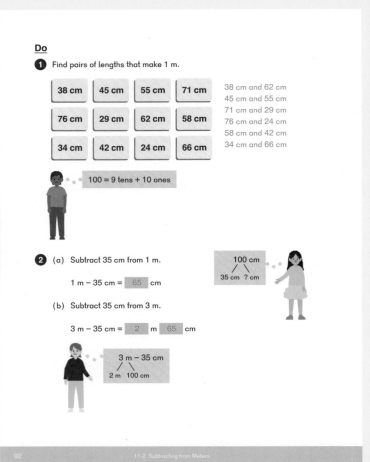

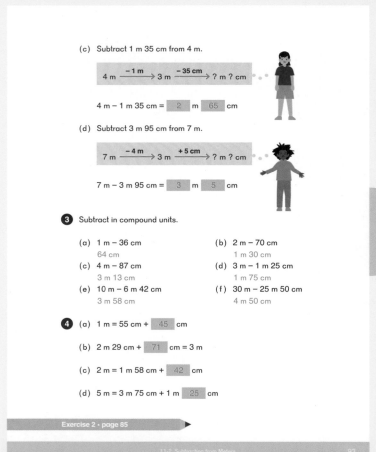

110　　　Teacher's Guide 3B Chapter 11　　　© 2017 Singapore Math Inc.

Lesson 3 Kilometers

Objectives
- Understand and use the kilometer as a unit to measure long distances.
- Convert between kilometers and meters.
- Distinguish between travel distance and direct distance.

Think

Pose the **Think** problems and discuss student solutions. Students will use meters for their solutions and will be introduced to kilometers in **Learn**.

Learn

(a) Alex introduces kilometers as a measure of length. 1,000 m = 1 km.

Discuss distances that might be measured in kilometers. Students may be familiar with miles. A kilometer is just over $1\frac{1}{2}$ miles.

Have students share when kilometers might be used instead of centimeters or meters. (Height of a mountain, drive to another city, how far a plane flies, etc.) Ensure that students understand that a kilometer is a distance we cannot measure with measuring tools in the classroom.

(b) We can express distances over 1,000 meters in compound units.

Dion splits 1,300 m into 1,000 m and 300 m. He knows 1,000 m = 1 km.

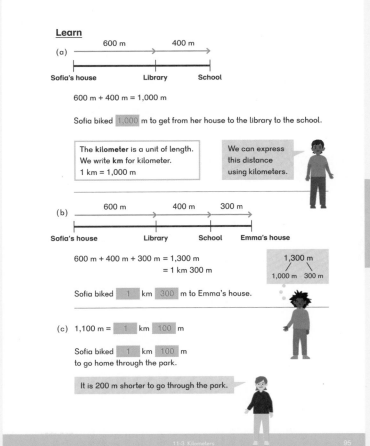

© 2017 Singapore Math Inc.   Teacher's Guide 3B Chapter 11   111

Do

① Distances traveled are rarely in a perfectly straight line. Dion and Emma introduce the terms "travel distance" and "direct distance."

In word problems, "travel distance" is typically used.

③ Check that students understand that:

4 km 50 m = 4,000 m + 50 m = 4,050 m and not 450 m or 4,500 m.

A common student error is to forget that there are 1,000 meters in a kilometer, whereas there are 100 centimeters in a meter.

Activity

▲ Estimate Length

Find a school track or measure the distance around the playground. Have students first predict how long it will take to walk or jog 1 km around the track. Then have them do it. Have them count how many laps and time how long it takes. Note that it typically takes 15 minutes to walk 1 km at a walking pace.

Exercise 3 • page 88

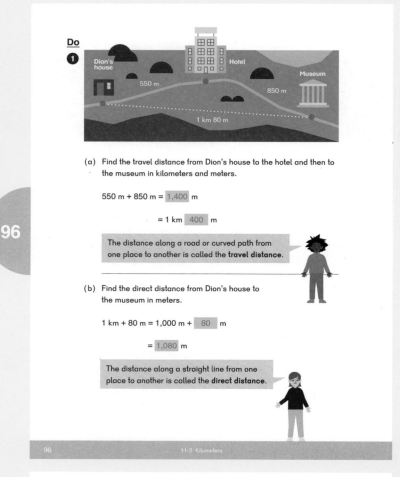

Lesson 4 Subtracting from Kilometers

Objective

- Subtract meters or kilometers and meters from a whole number of kilometers.

Think

Pose the **Think** problem and have students work independently. Discuss student solutions. Ask students how subtracting from a kilometer is similar to or different from subtracting centimeters.

Learn

Method 1

Mei splits 2 km into 1 km and 1,000 m. She subtracts 750 m from 1,000 m and adds back in the 1 km.

Method 2

Dion over-subtracts. He subtracts 1 km and adds back in 250 m.

Method 3

Emma converts 2 km to 2,000 m. She subtracts 750 m from 2,000 m, then converts 1,250 m to kilometers and meters.

Have students compare their methods from **Think** with the ones in the textbook.

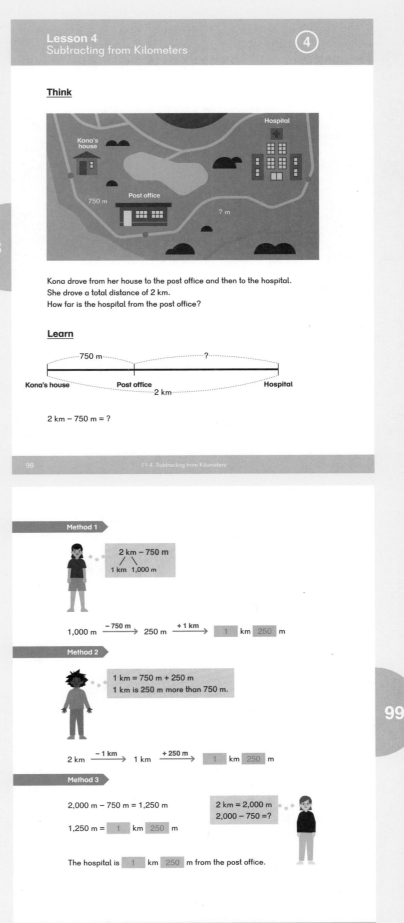

© 2017 Singapore Math Inc.   Teacher's Guide 3B Chapter 11   113

Do

1. This problem could be used as an activity by writing these pairs and similar sums to 1,000 m on index cards.

2. (b) Dion knows 3 km is 2 km 1,000 m. He subtracts 610 from 1,000 m, then adds back the 2 km.

5—6 Students can use any strategy to find the answer.

Activity

▲ Estimate Distance

Have students estimate the distance in kilometers by car to locations near their home or school. Have them find actual distances in kilometers using the internet or a map.

Exercise 4 • page 91

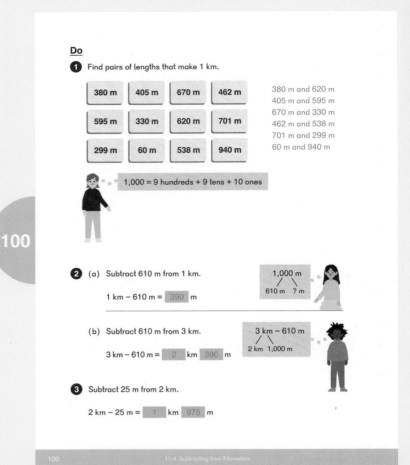

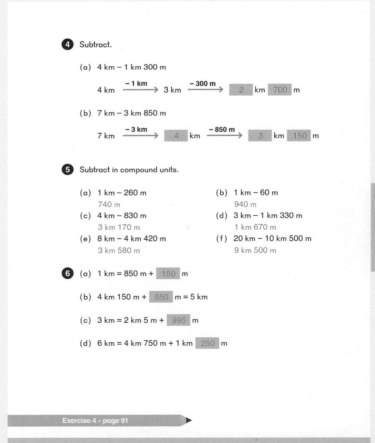

Lesson 5 Liters and Milliliters

Objectives

- Understand capacity and liquid volume.
- Measure capacity in liters and milliliters.

Lesson Materials

- Small paper cups
- 500 mL beakers
- Water bottles ($\frac{1}{2}$ liter)
- 1,000 mL beakers
- Water
- Teaspoons

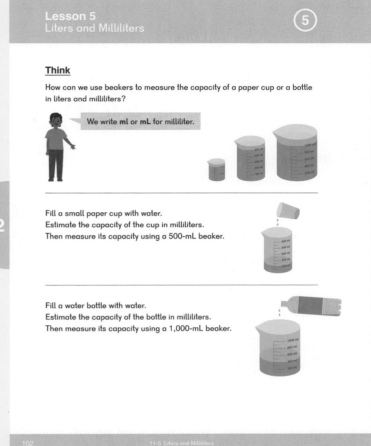

Think

Discuss the **Think** tasks and Alex's comment about milliliters. This is a new term for students.

Provide students with water, small cups, and beakers. Discuss the markings and the scale on the beakers and have students complete the task.

Students should note that 1 milliliter is a very small amount of water.

Note: ❷ should be completed at this time, while water and containers are out.

Ensure students have a feeling for how much a milliliter is. Have them estimate, then find the number of milliliters a small container will hold.

Learn

Discuss the beakers and Dion and Mei's comments. Just as we can use compound units to measure distance, we also can use compound units to measure capacity. Students should be familiar with liters. Introduce the abbreviation mL for milliliters.

Note that the tick marks vary on the beakers. Dion's beaker shows 50 mL increments, and only increments of 100 mL are labeled.

Mei's beaker shows 100 mL increments, and only increments of 200 mL are labeled.

Discuss the terms "capacity" and "volume" and how they are different.

Capacity is the amount of liquid a container can hold. The amount of liquid in the container is the volume of the liquid. When we use a beaker to measure the amount of liquid, that is not the capacity of the beaker itself, it is the amount of liquid in the container. A container can hold less liquid than its capacity.

Show students a teaspoon to give them an idea of what 5 mL of water is.

Do

Dion shares that "milli" comes from the Latin word meaning "thousand."

❶ Convert 1,000 mL to 1 L. When we divide 1 L into 1,000 equal units, 1 unit is 1 mL.

❷ Students should have completed this task during the **Think** portion of the lesson.

❸ The beaker is filled halfway between 200 mL and 300 mL.

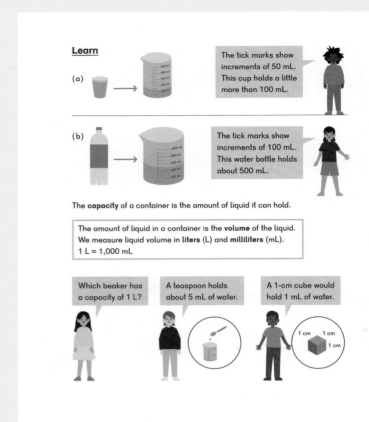

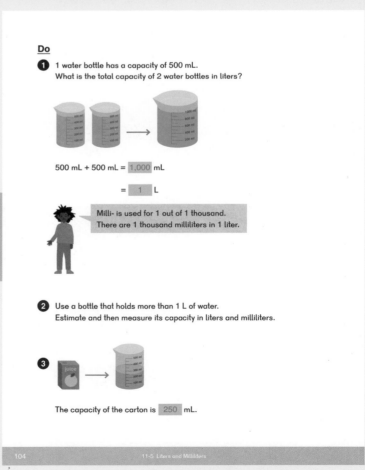

116   Teacher's Guide 3B Chapter 11   © 2017 Singapore Math Inc.

5 Sofia reminds students:

1,000 mL = 1 L
4 × 1,000 mL = 4,000 mL = 4 L

6 Verify students understand that:

2 L 50 mL = 2,000 mL + 50 mL = 2,050 mL, and not 250 mL or 2,500 mL.

9 (b) You would need to fill the partially-full beaker with 150 mL of water and fill another 500 mL beaker with water to have 2 L.

500 mL + 500 mL = 1L
350 mL + 150 mL + 500 mL = 1L

Exercise 5 • page 94

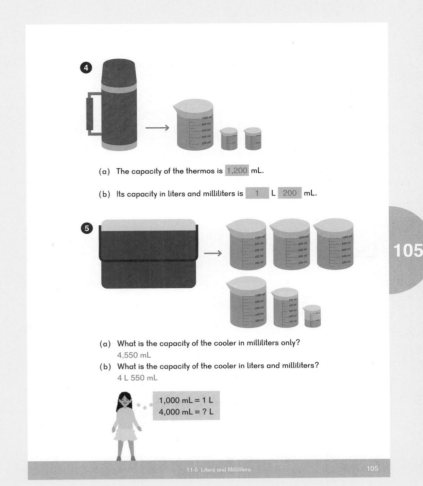

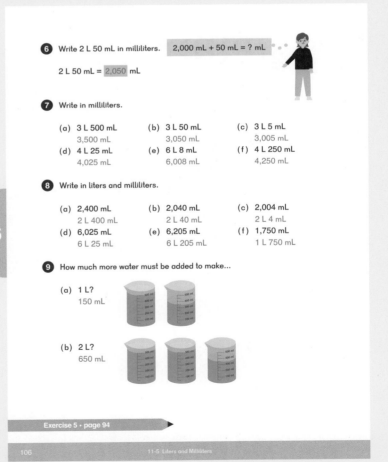

Lesson 6 Kilograms and Grams

Objectives

- Understand weight and mass.
- Measure mass in grams and kilograms.

Lesson Materials

- Platform scales measuring in metric units
- Items that weigh between 1 kg and 5 kg
- Weight sets

Think

Discuss the **Think** tasks and Sofia's comment about kilograms. Provide students with platform scales and discuss the markings on the scales. Hand out items to weigh and have students complete the task.

By this lesson, it is expected that students can see that weight can also be expressed in compound units (kilograms and grams). Both grams and kilograms were introduced in Dimensions Math 2A Chapter 5: Weight.

Learn

Alex shares that "kilo" comes from the Greek word meaning "thousand."

Some countries use the term "kilo" as a short form of kilogram only. Kilo is a prefix on many words such as "kilometer," "kilowatt," etc. One might say, "I weigh 35 kilos," but it is not customary to say, "I walked 3 kilos."

Sofia points out that 1 mL weighs 1 g. Note that a plastic 1 centimeter cube also weighs 1 gram.

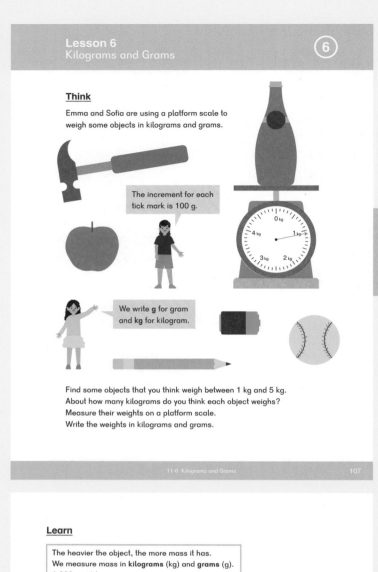

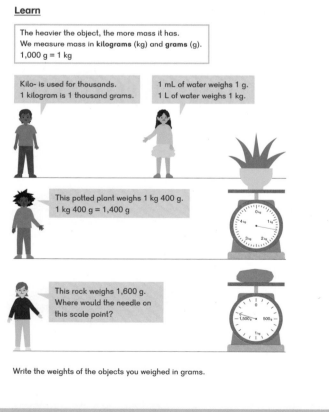

Do

1. Students need platform scales and weights.

2. Note that the tick marks are different on the two scales. Students should see that on each of these scales, the intervals denote 100 grams.

3. Verify students understand that:

    3 kg 45 g = 3,000 g + 45 g = 3,045 g, and not 345 g or 3,450 g.

Exercise 6 • page 97

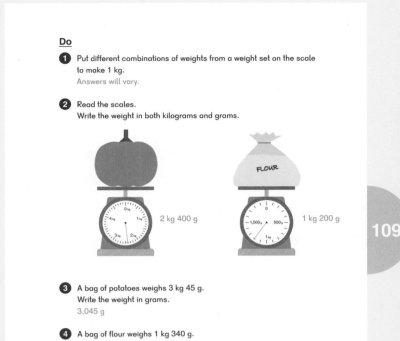

Lesson 7 Word Problems

Objective

- Solve word problems involving measurement.

Think

Pose the **Think** problem and discuss the bar model. Have students share their solutions to the problem.

Dion provides a hint. Students should find the distance between the hotel and Alex's house first.

Discuss student solutions.

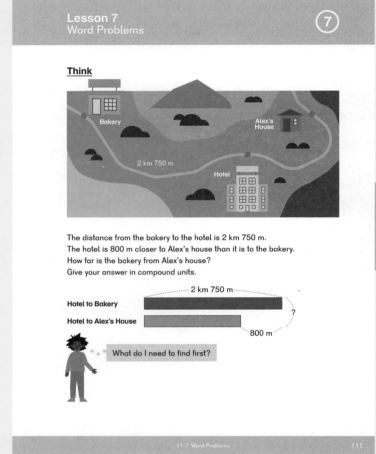

Learn

Emma thinks of 2 km 750 m as 2,750 m. Dion then uses the vertical algorithm to subtract 800 from 2,750.

Students could also use a different strategy.

For example:

```
2 km 750 m − 800 m =
      /    \
    750    50

2 km − 50 m = 1 km 950 m
```

Students could then add:

```
2 km 750 m + 1 km 950 m =
      /    \
    700    50

3 km + 1 km + 700 m = 4 km 700 m
```

Or:

```
  2,750
+ 1,950
  4,700    4 km 700 m
```

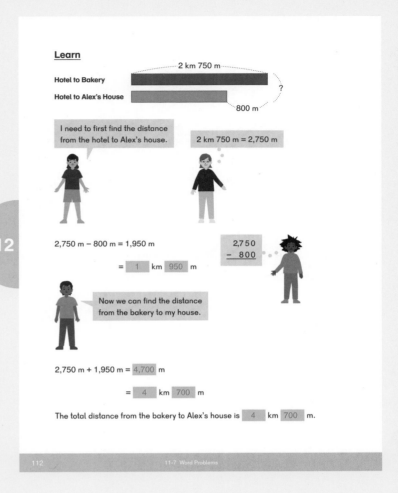

Have students compare their methods from **Think** with the ones in the textbook.

Students can also solve the problem using equations, with a letter standing for the unknown quantity:

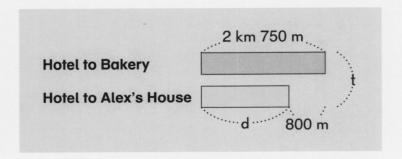

d = distance from Hotel to Alex's House
t = distance from Bakery to Alex's House

d = 2,750 − 800
t = d + 2,750

Do

3 Sample two-step bar model:

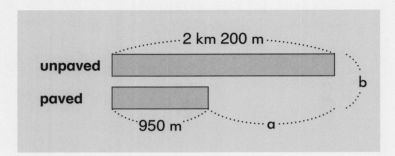

Students can also solve the problem using equations, with a letter standing for the unknown quantity:

a = difference between paved and unpaved trail
b = distance of the paved and unpaved trails combined

a = 2,200 − 950
b = 2,200 + 950

4 Mei converts to grams, then uses a vertical algorithm.

Do

1 Dion has a board that is 2 m 20 cm long.
He cuts off 90 cm to make a ramp for his remote control car.

(a) What is the length of the remaining piece of board?

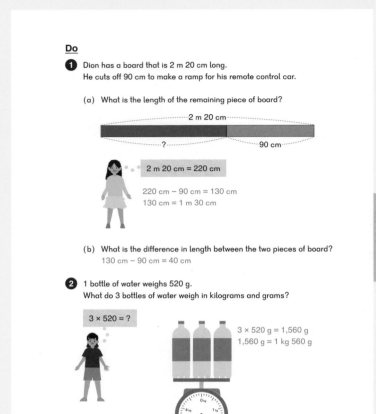

2 m 20 cm = 220 cm

220 cm − 90 cm = 130 cm
130 cm = 1 m 30 cm

(b) What is the difference in length between the two pieces of board?
130 cm − 90 cm = 40 cm

2 1 bottle of water weighs 520 g.
What do 3 bottles of water weigh in kilograms and grams?

3 × 520 = ?

3 × 520 g = 1,560 g
1,560 g = 1 kg 560 g

3 There are two trails to the scenic overlook.
The paved trail is 950 m.
The unpaved trail is 2 km 200 m.

(a) How much farther is it to walk on the unpaved trail than on the paved trail in kilometers and meters?
2,200 m − 950 m = 1,250 m; 1 km 250 m

(b) Sofia went to the scenic overlook on the unpaved trail, and came back on the paved trail.
How far did she walk?
2,200 m + 950 m = 3,150 m
3 km 150 m

4 How much do the can of paint and the bucket of plaster weigh altogether?

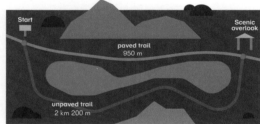

9 kg 615 g   5 kg 930 g   3,685 g

6

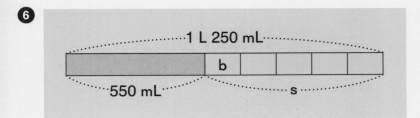

Students can also solve the problem using equations, with a letter standing for the unknown quantity:

s = amount of solution poured into the 5 beakers
b = amount of solution in 1 beaker

s = 1,250 − 550
b = s ÷ 5

Exercise 7 • page 101

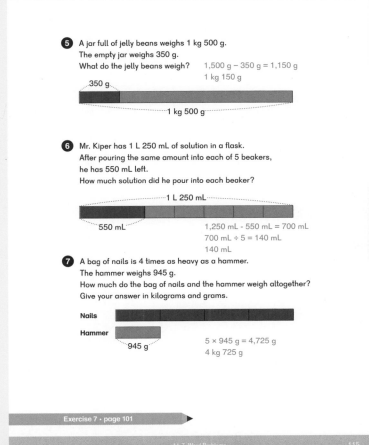

Lesson 8 Practice

Objective

- Practice adding and subtracting compound units of measurement.

After students complete the **Practice** in the textbook, have them continue to practice adding and subtracting measurement with activities from the chapter.

Activities

Have students write their own measurement problems and exchange with classmates to solve them.

▲ **Match**

Materials: Measurement Match Cards (BLM)

Use **Match** and **Memory** to review measurements or equivalent measurements.

Lay cards in a faceup array. Have students match two Measurement Match Cards (BLM) that show equivalent measurements.

▲ **Memory**

Materials: Measurement Match Cards (BLM)

Play using the same rules as **Match**, but set the cards out facedown in an array.

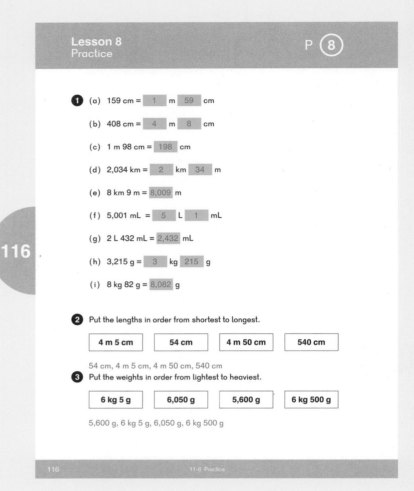

11

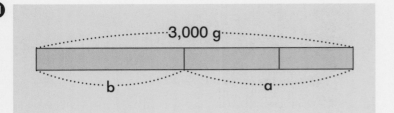

Students can also solve the problem using equations, with a letter standing for the unknown quantity:

a = amount of flour used
b = amount of flour left

a = 900 + 700
b = 3,000 − a

Brain Works

★ **Fruit Combinations**

What combination of fruit weighs the same as the apple?

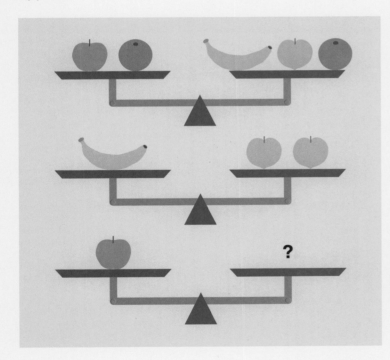

Answer:

3 peaches or a banana and a peach

Exercise 8 • page 105

---

4 (a) 1 m − 73 cm = 27 cm
(b) 2 m − 9 cm = 1 m 91 cm
(c) 3 m − 1 m 5 cm = 1 m 95 cm
(d) 1 km − 620 m = 380 m
(e) 4 km − 90 m = 3 km 910 m
(f) 5 km − 2 km 850 m = 2 km 150 m
(g) 1 L − 405 mL = 595 mL
(h) 4 L − 35 mL = 3 L 965 mL
(i) 6 kg − 120 g = 5 kg 880 g
(j) 7 kg − 5 g = 6 kg 995 g

5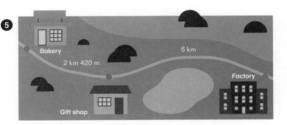

The gift shop is how much closer to the bakery than to the factory?
6 km − 2 km 420 m = 3 km 580 m

6 The capacity of a canteen is 1 L 500 mL.
The capacity of a thermos is 655 mL.

(a) What is the total capacity of the canteen and thermos in liters and milliliters?
1,500 mL + 655 mL = 2,155 mL; 2 L 155 mL

(b) How much more water can the canteen hold than the thermos?
1,500 mL − 655 mL = 845 mL
845 mL

7 Sebastian is 1 m 35 cm tall.
His brother is 96 cm tall.
Who is taller and by how much?
135 cm − 96 cm = 39 cm
Sebastian, 39 cm taller

8 A plank of wood is 5 m 85 cm long.
It is cut into 9 pieces of equal length.
How long is each piece?
585 cm ÷ 9 = 65 cm
65 cm

9 Hudson walked a total of 950 m along the river trail and then back again.
How far did he walk in kilometers and meters?
950 m × 2 = 1,900 m
1 km 900 m

10 Susma ran around a 515-m track 4 times.
How far did she run in kilometers and meters?
515 m × 4 = 2,060 m
2 km 60 m

11 A baker has 3 kg of flour.
She uses 900 g to make muffins and 700 g to make pastries.
How many kilograms and grams of flour does she have left?
900 g + 700 g = 1,600 g = 1 kg 600 g
3 kg − 1 kg 600 g = 1 kg 400 g

Exercise 8 • page 105

Review 3

Objective

- Review topics from Chapter 1 through Chapter 11.

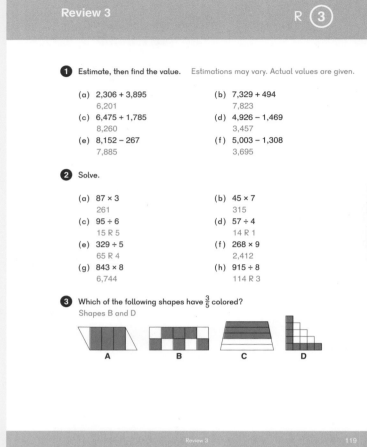

Use this cumulative review as needed to practice and reinforce content and skills from the first 11 chapters.

Brain Works

★ Problems Without Numbers

Often when students encounter a word problem, they focus on computing some quantity rather than what is being asked. These numberless problems can be a good way to get students to think about the relevant underlying mathematical ideas.

(a) I know the length and width of a room and the number of square meters that a can of paint will cover. How can I find how many cans of paint I will need to buy to paint the walls of that room?

(b) How can I find how many times a bicycle wheel will turn going 2 kilometers?

(c) How can I find the distance around a square dog park if I know how long two-thirds of one side is?

(d) I know how many centimeters long and wide a tile is, and how many meters long and wide a room is. How do I find how many tiles will cover the floor?

(e) I know how many milliliters of juice it takes to fill a small pitcher. How can I find how many liters it takes to fill 5 such pitchers?

(f) I know how much my dog weighs when standing on four feet. How do I find his weight when standing on three feet?

(g) I know the weight of a box of gumballs in bags and the weight of the empty box. What else do I need to know to find the weight of each bag of gumballs in the box?

4)
```
0   A   B   C   1   D   E
```

(a) What number is indicated by C?
   $\frac{3}{4}$
(b) Which letter indicates the number $\frac{5}{4}$?
   D
(c) How many eighths make the number indicated by B?
   4 eighths

5) What sign, >, <, or =, goes in the ◯?

(a) $\frac{5}{6} \bigcirc{>} \frac{5}{8}$ (b) $\frac{4}{8} \bigcirc{>} \frac{7}{16}$

(c) $\frac{3}{5} \bigcirc{>} \frac{1}{3}$ (d) $\frac{2}{3} \bigcirc{>} \frac{4}{7}$

(e) $\frac{3}{7} + \frac{2}{7} \bigcirc{<} \frac{7}{8} - \frac{2}{8}$ (f) $\frac{11}{12} - \frac{5}{12} \bigcirc{<} \frac{3}{8} + \frac{3}{8}$

6) (a) The volume of water in this beaker is $\boxed{2}$ L $\boxed{500}$ mL.

(b) 750 mL of water is poured out. How much water is still in the beaker? Give your answer in compound units.
   2,500 mL − 750 mL = 1,750 mL
   1 L 750 mL

7) This bar graph shows the length of five different hikes in kilometers.

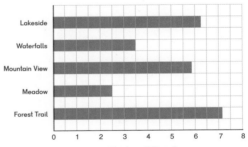

(a) Each square on the graph shows increments of $\boxed{500}$ m.
(b) List the hikes in order from shortest to longest.
   Meadow, Waterfalls, Mountain View, Lakeside, Forest Trail
(c) Which two hikes are about the same length?
   Mountain View and Lakeside
(d) Dion went on the hike that was 6 km 200 m. Which hike did he go on?
   Lakeside
(e) Sofia went on two hikes. She hiked about 8 km. Which two hikes did she go on?
   Mountain View and Meadow

9 7 units ⟶ 98 seashells
1 unit ⟶ 98 ÷ 7 = 14 seashells
2 units ⟶ 2 × 14 = 28 seashells

10 By adding 200 to the second number, students make two bars of equal lengths. They can divide by 2 to find the greater number.

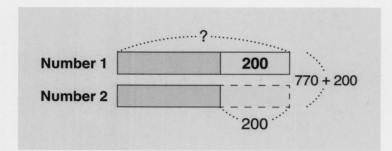

11 Ask students if they can find the weight of 1 box and then 1 crate first.

Exercise 9 • page 109

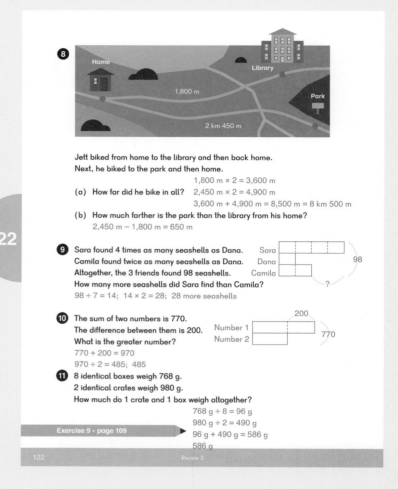

Exercise 1 • pages 81–84

Chapter 11 Measurement

Exercise 1

Basics

1. Fill in the blanks with m for meters or cm for centimeters.

   (a) A baseball bat is about 1 __m__ long.

   (b) An 8-year-old child is about 125 __cm__ tall.

   (c) A 3-story building is about 10 __m__ tall.

   (d) A hummingbird is about 10 __cm__ long.

2. (a) 1 m = 100 cm   (b) 100 cm = __1__ m

   2 m = __200__ cm   200 cm = __2__ m

   6 m = __600__ cm   500 cm = __5__ m

   10 m = __1,000__ cm   900 cm = __9__ m

3. A van is 8 m 15 cm long. Find its length in centimeters.

   8 m 15 cm = 800 cm + __15__ cm

   = __815__ cm

   The van is __815__ cm long.

4. A flag pole has a height of 740 cm. Find its height in meters and centimeters.

   740 cm = 700 cm + __40__ cm

   = 7 m __40__ cm

   The flag pole has a height of __7__ m __40__ cm.

5. A bed is 203 cm long. Find its length in meters and centimeters.

   203 cm = 200 cm + __3__ cm

   = __2__ m __3__ cm

   The bed is __2__ m __3__ cm long.

6. Write the lengths in order, from shortest to longest.

   3 m 2 cm, 203 cm, 320 cm, 2 m 30 cm

   203 cm, 2 m 30 cm, 3 m 2 cm, 320 cm

Practice

7. (a) 5 m 30 cm = __530__ cm

   (b) 6 m 8 cm = __608__ cm

   (c) 2 m 35 cm = __235__ cm

   (d) 7 m 27 cm = __727__ cm

8. (a) 870 cm = __8__ m __70__ cm

   (b) 525 cm = __5__ m __25__ cm

   (c) 602 cm = __6__ m __2__ cm

9. (a) Circle the best estimate for the length of the computer monitor.

   10 cm   **(60 cm)**

   200 cm   1 m 5 cm

   (b) Circle the best estimate for the height of the tree.

   75 cm   1 m 30 cm

   4 m   **(2 m 90 cm)**

   200 cm   105 cm

   (c) Circle the best estimate for length of the car.

   700 cm   4 cm

   6 m   45 cm

   (4 m 50 cm)   6 m 20 cm

10. Write >, <, or = in each ◯.

    (a) 4 m 25 cm **>** 245 cm

    (b) 101 cm **=** 1 m 1 cm

    (c) 570 cm **>** 5 m 7 cm

    (d) 8 m 18 cm **<** 881 cm

11. Two sofas are put side by side with a gap of 5 cm between them.
    The total length is 261 cm.
    How long is one sofa in meters and centimeters?
    261 cm − 5 cm = 256 cm
    256 cm ÷ 2 = 128 cm
    128 cm = 1 m 28 cm

Challenge

12. (a) 10 m 25 cm = __1,025__ cm

    (b) 6,185 cm = __61__ m __85__ cm

13. A stack of 8 blocks has a height of 40 cm.
    How many blocks are needed to make a stack that is 2 m high?
    40 cm ÷ 8 = 5 cm
    Each block has a height of 5 cm.
    2 m = 200 cm
    200 cm ÷ 5 = 40
    40 blocks are needed to make a stack that is 2 m high.

Exercise 2 • pages 85–87

Exercise 2

Basics

1. Draw straight lines to join each pair of lengths that add up to 1 m. You will get 5 triangles.

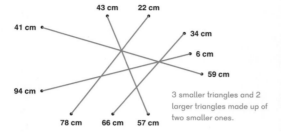

3 smaller triangles and 2 larger triangles made up of two smaller ones.

2. (a) 1 m − 41 cm = **59** cm   2 m − 41 cm = 1 m **59** cm
   (b) 1 m − 22 cm = **78** cm   5 m − 22 cm = 4 m **78** cm

3. (a) 1 m − 6 cm = **94** cm
   (b) 6 m − 6 cm = **5** m **94** cm
   (c) 1 m − 66 cm = **34** cm
   (d) 10 m − 66 cm = **9** m **34** cm
   (e) 15 m − 78 cm = **14** m **22** cm

4. (a) 1 m − 43 cm = **57** cm
   (b) 7 m − 2 m = **5** m
   (c) 5 m − 43 cm = 4 m **57** cm
   (d) 7 m − 2 m 43 cm = 4 m **57** cm

Practice

5. (a) 10 m − 8 m = **2** m
   (b) 2 m − 98 cm = **1** m **2** cm
   (c) 10 m − 8 m 98 cm = **1** m **2** cm

6. (a) 1 m − 75 cm = **25** cm
   (b) 1 m − 82 cm = **18** cm
   (c) 2 m − 35 cm = **1** m **65** cm
   (d) 2 m − 82 cm = **1** m **18** cm
   (e) 8 m − 3 cm = **7** m **97** cm

7. (a) 35 cm + **65** cm = 1 m   (b) 86 cm + **14** cm = 1 m
   (c) 8 cm + **92** cm = 1 m   (d) **72** cm + 28 cm = 1 m

8. (a) 6 m − 1 m 20 cm = **4** m **80** cm
   (b) 9 m − 5 m 93 cm = **3** m **7** cm
   (c) 4 m − 2 m 6 cm = **1** m **94** cm
   (d) 9 m − 8 m 91 cm = **9** cm

9. (a) 2 m 29 cm + **71** cm = 3 m
   (b) 5 m 62 cm + **38** cm = 6 m
   (c) 7 m **82** cm + 18 cm = 8 m
   (d) 1 m **63** cm + 37 cm = 2 m

10. Rope A is 5 m 48 cm long and Rope B is 8 m long.
    What is the difference in length between the two ropes?
    8 m − 5 m 48 cm = 2 m 52 cm
    The difference in length between the two ropes is 2 m 52 cm.

11. A pole that is 7 m long is painted in two colors.
    The first 3 m is painted green, and the last 65 cm is also painted green.
    The middle portion is painted yellow.
    What length of the pole is painted yellow?
    7 m − 3 m 65 cm = 3 m 35 cm
    3 m 35 cm of the pole is painted yellow.

Exercise 3 • pages 88–90

Exercise 3

Basics

1. Fill in the blanks with km for kilometers or m for meters.

   (a) A bus is about 10 __m__ long.

   (b) Shanice went on a 6 __km__ hike.

   (c) The distance from Los Angeles to San Diego is about 180 __km__.

   (d) Mount Blackburn in Alaska is just under 5 __km__ high.

   (e) The Grand Canyon reaches a depth of about 1,850 __m__.

2. (a) 1 km = 1,000 m   (b) 1,000 m = 1 km

   2 km = 2,000 m   2,000 m = 2 km

   5 km = 5,000 m   6,000 m = 6 km

   7 km = 7,000 m   9,000 m = 9 km

3. The height of Mount Everest is 8 km 848 m.
   Write its height in meters.

   8 km 848 m = 8,000 m + 848 m
             = 8,848 m

   Mount Everest has a height of __8,848__ m.

4. The Hayes Volcano in Alaska has a height of 3,034 m above sea level.
   Find its height in kilometers and meters.

   3,034 m = 3,000 m + 34 m

   = 3 km 34 m

   Hayes is __3__ km __34__ m high.

Practice

5. (a) 5 km 300 m = 5,300 m   (b) 2 km 205 m = 2,205 m

   (c) 9 km 819 m = 9,819 m   (d) 6 km 80 m = 6,080 m

   (e) 1 km 10 m = 1,010 m   (f) 7 km 7 m = 7,007 m

6. (a) 8,700 m = 8 km 700 m

   (b) 9,147 m = 9 km 147 m

   (c) 5,065 m = 5 km 65 m

   (d) 6,002 m = 6 km 2 m

7. Write >, <, or = in each ◯.

   (a) 2 km 520 m > 2,450 m   (b) 8 km 18 m > 818 m

   (c) 5,100 m > 5 km 1 m   (d) 5,070 m = 5 km 70 m

8. This map shows the length of some trails near a campground.

   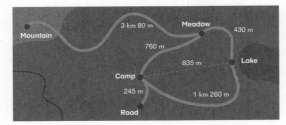

   (a) What is the direct distance from the camp to the lake?
   835 m

   (b) What is the distance from the meadow to the mountain in meters?
   3,080 m

   (c) What is the total distance from the road to the meadow, passing through the camp, in kilometers and meters?
   245 m + 760 m = 1,005 m = 1 km 5 m

   (d) What is the shortest distance from the camp to the lake along the trails in kilometers and meters?
   760 m + 430 m = 1,190 m
   1 km 190 m
   1,190 m < 1 km 260 m

Exercise 4 • pages 91–93

Exercise 4

Basics

1. (a) 1 km = 900 m + 90 m + [10] m

   (b) 700 m + [200] m = 900 m

   30 m + [60] m = 90 m

   4 m + [6] m = 10 m

   734 m + [266] m = 1 km

2. Draw straight lines to join each pair of lengths that add up to 1 km. You will get 2 pentagons and 1 triangle.

   1 pentagon consists of the smaller pentagon and the triangle.

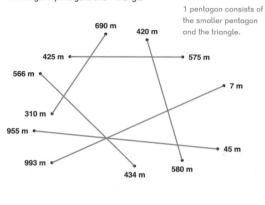

3. (a) 1 km – 690 m = [310] m | 2 km – 690 m = 1 km [310] m

   (b) 1 km – 993 m = [7] m | 5 km – 993 m = 4 km [7] m

4. (a) 1 km – 420 m = [580] m

   (b) 3 km – 420 m = [2] km [580] m

   (c) 9 km – 6 km = [3] km

   (d) 9 km – 6 km 420 m = [2] km [580] m

Practice

5. (a) 1 km – 650 m = [350] m   (b) 1 km – 652 m = [348] m

   (c) 1 km – 9 m = [991] m   (d) 1 km – 27 m = [973] m

   (e) 1 km – 349 m = [651] m

6. (a) 2 km – 349 m = 1 km [651] m

   (b) 2 km – 1 km 700 m = [300] m

   (c) 2 km – 1 km 50 m = [950] m

   (d) 3 km – 1 km 5 m = [1] km [995] m

   (e) 6 km – 3 km 470 m = [2] km [530] m

   (f) 10 km – 3 km 47 m = [6] km [953] m

7. This map shows the length of some trails near a campground.

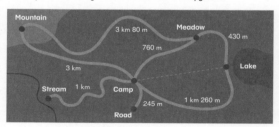

Find the differences in the length for the following trails, using the shortest possible routes.
Give your answers in compound units when possible.

(a) The camp to the stream and the meadow to the lake.
1 km – 430 m = 570 m

(b) The camp to the stream and the camp to the road.
1 km – 245 m = 755 m

(c) The camp to the mountain and the camp to the meadow.
3 km – 760 m = 2 km 240 m

(d) The camp to the mountain and the camp to the lake.
3 km – 1 km 260 m = 1 km 740 m

(e) The road to the meadow through the camp, and the camp to the mountain.
245 m + 760 m = 1,005 m = 1 km 5 m
3 km – 1 km 5 m = 1 km 995 m

Exercise 5 • pages 94–96

Exercise 5

Basics

1 Fill in the blanks with L for liters or mL for milliliters.

(a) The capacity of a soda bottle is 1 __L__.

(b) A 1-cm cube could hold 1 __mL__ of water.

(c) A tablespoon holds about 15 __mL__ of water.

(d) A bottle of cooking oil holds about 1 __L__ of oil.

(e) The capacity of a teacup is about 250 __mL__.

2 (a) 1 L = 1,000 milliliters (b) 1,000 mL = __1__ L

3 L = __3,000__ mL   6,000 mL = __6__ L

3 How much water is in each beaker?

(a) 700 mL

(b) 300 mL

(c) 350 mL

(a) Find the capacity of this container in milliliters.

1,000 mL + 1,000 mL + 250 mL = __2,250__ mL

(b) Find the capacity of this container in liters and milliliters.

1 L + 1 L + 250 mL = __2__ L __250__ mL

5 (a) 4,000 mL = __4__ L

(b) 4,200 mL = __4__ L __200__ mL

(c) 4,020 mL = __4__ L __20__ mL

(d) 4,002 mL = __4__ L __2__ mL

6 (a) 1 L − 230 mL = __770__ mL

(b) 2 L − 230 mL = __1__ L __770__ mL

(c) 8 L − 4 L 230 mL = __3__ L __770__ mL

Practice

7 (a) 5 L 734 mL = __5,734__ mL

(b) 8 L 32 mL = __8,032__ mL

(c) 3,705 mL = __3__ L __705__ mL

(d) 6,043 mL = __6__ L __43__ mL

8 (a) 1 L − 345 mL = __655__ mL

(b) 7 L − 6 L 45 mL = __955__ mL

(c) 3 L − 3 mL = __2__ L __997__ mL

(d) 7 L − 2 L 380 mL = __4__ L __620__ mL

(e) 9 L − 4 L 75 mL = __4__ L __925__ mL

Challenge

9 How can we put 4 L in the bucket using only the 5-L and 3-L containers?

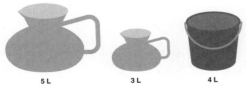

5 L   3 L   4 L

Fill the 5-L container and use it to fill the 3-L container completely.
Pour the remaining 2 L from the 5-L container into the bucket.
Repeat one time and there will be 4 L in the bucket.

Teacher's Guide 3B Chapter 11

Exercise 6 • pages 97–100

Exercise 6

Basics

1. Fill in the blanks with kg for kilograms or g for grams.

   (a) A paper clip weighs about 1 __g__.

   (b) A liter of water weighs 1 __kg__.

   (c) An 8-year-old child weighs about 30 __kg__.

   (d) A nickel weighs about 6 __g__.

   (e) An elephant can weigh 4,500 __kg__.

   (f) A cat can weigh 4,500 __g__.

2. (a) 1 kg = 1,000 grams   (b) 1,000 g = __1__ kg

   4 kg = __4,000__ g   7,000 g = __7__ kg

3. (a) 6 kg 260 g = __6,260__ g   (b) 6 kg 26 g = __6,026__ g

4. (a) 4,900 g = __4__ kg __900__ g

   (b) 4,009 g = __4__ kg __9__ g

5. (a) 1 kg − 430 g = __570__ g   (b) 8 kg − 430 g = 7 kg __570__ g

6. Write the weights of the following items.

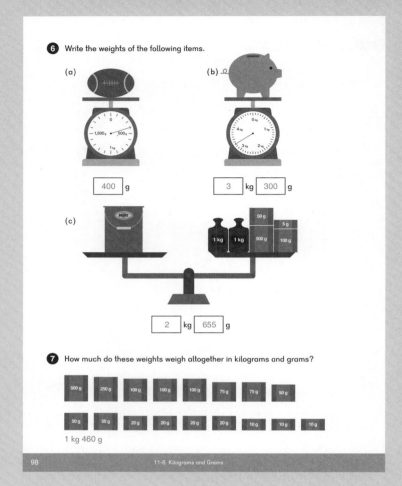

   (a) 400 g

   (b) 3 kg 300 g

   (c) 2 kg 655 g

7. How much do these weights weigh altogether in kilograms and grams?

   1 kg 460 g

Practice

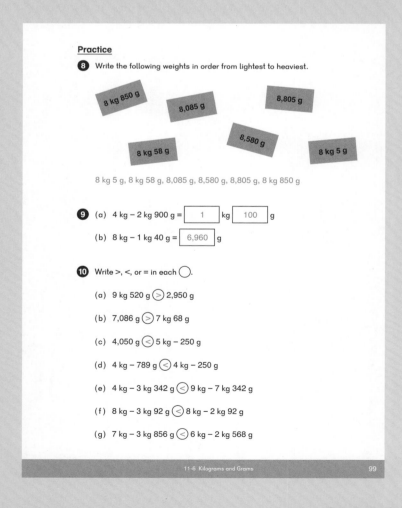

8. Write the following weights in order from lightest to heaviest.

   8 kg 850 g, 8,085 g, 8,805 g, 8 kg 58 g, 8,580 g, 8 kg 5 g

   8 kg 5 g, 8 kg 58 g, 8,085 g, 8,580 g, 8,805 g, 8 kg 850 g

9. (a) 4 kg − 2 kg 900 g = __1__ kg __100__ g

   (b) 8 kg − 1 kg 40 g = __6,960__ g

10. Write >, <, or = in each ◯.

    (a) 9 kg 520 g __>__ 2,950 g

    (b) 7,086 g __>__ 7 kg 68 g

    (c) 4,050 g __<__ 5 kg − 250 g

    (d) 4 kg − 789 g __<__ 4 kg − 250 g

    (e) 4 kg − 3 kg 342 g __<__ 9 kg − 7 kg 342 g

    (f) 8 kg − 3 kg 92 g __<__ 8 kg − 2 kg 92 g

    (g) 7 kg − 3 kg 856 g __<__ 6 kg − 2 kg 568 g

Challenge

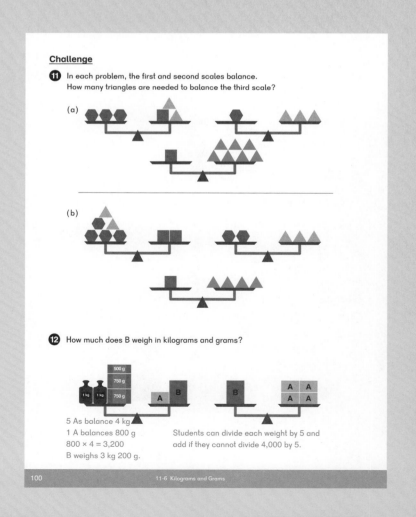

11. In each problem, the first and second scales balance. How many triangles are needed to balance the third scale?

    (a)

    (b)

12. How much does B weigh in kilograms and grams?

    5 As balance 4 kg
    1 A balances 800 g
    800 × 4 = 3,200
    B weighs 3 kg 200 g.

    Students can divide each weight by 5 and add if they cannot divide 4,000 by 5.

Exercise 7 • pages 101–104

Exercise 7

Basics

1 A tank can hold 4 L 300 mL of water.
It has 2 L 450 mL of water in it now.
How much more water is needed to fill it?

4 L 300 mL − 2 L 450 mL = | 4,300 | mL − | 2,450 | mL

= | 1,850 | mL

= | 1 | L | 850 | mL

__1 L 850 mL__ more water is needed to fill the tank.

2 1 bag of flour weighs 850 g.
How much do 6 such bags of flour weigh?

850 g × 6 = | 5,100 | g = | 5 | kg | 100 | g

6 bags of flour weigh __5 kg 100 g__.

3 9 bricks placed end to end are 2 m 7 cm long.
How long is 1 brick?

2 m 7 cm = | 207 | cm

| 207 | cm ÷ 9 = | 23 | cm

1 brick is __23 cm__ long.

4 The total weight of 3 tennis balls and 8 marbles is 380 g.
Each tennis ball weighs 60 g.
How much does 1 marble weigh?

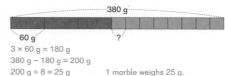

3 × 60 g = 180 g
380 g − 180 g = 200 g
200 g ÷ 8 = 25 g   1 marble weighs 25 g.

Practice

5 This map shows the length of some trails near a campground.
Write any answers greater than 1 km in compound units.
Assume the problem is asking for the shortest possible distances.

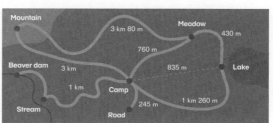

(a) How much longer is the distance along the trail from the camp to the lake through the meadow than the direct distance to the lake?
760 m + 430 m = 1,190 m (distance from the camp to the lake)
1,190 m − 835 m = 355 m
It is 355 m longer.

(b) What is the total distance of the loop from the camp to the lake, then to the meadow, then to the mountain, and then back to the camp?
Students may add km and m separately, or convert all to m first.
1 km + 3 km + 3 km = 7 km
260 m + 430 m + 80 m = 770 m
The total distance is 7 km 770 m.

(c) Abigail went from the camp to the meadow and back twice.
How far did she hike?
760 m × 4 = 3,040 m = 3 km 40 m
She hiked 3 km 40 m.

(d) Eli hiked from the camp past the lake to the meadow and then directly back to camp.
How much farther did Abigail hike than Eli?
1,260 m + 430 m + 760 m = 2,450 m
3,040 m − 2,450 m = 590 m
Abigail hiked 590 m farther than Eli.

Students may convert to all meters, use mental math to add or subtract within a kilometer, or a combination of these two methods.

(e) Maya hiked the trails from the camp to the beaver dam and back.
She hiked a total of 2 km 690 m.
How long is the trail from the stream to the beaver dam?

2 km 690 m − 1 km − 1 km = 690 m
690 m ÷ 2 = 345 m
The trail from the stream to the beaver dam is 345 m.

6 Santino cuts a 2-m rope into 8 equal pieces.
How long is each piece?
200 cm ÷ 8 = 25 cm
Each piece is 25 cm long.

7 A container has a total capacity of 3 L 90 mL.
It was filled with 2 L 455 mL of water.
970 mL more water was poured into it, with the extra water overflowing.
How much water overflowed?

2,455 mL + 970 mL = 3,425 mL
3,425 mL − 3,090 mL = 335 mL
335 mL of water overflowed.

Challenge

8 There are two types of weights, A and B.
3 of weight A and 2 of weight B weigh 1 kg 300 g altogether.
1 of weight A and 2 of weight B weigh 840 g altogether.
How much do 1 of weight A and 1 of weight B weigh altogether?

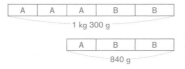

1,300 g − 840 g = 460 g (two As)
460 g ÷ 2 = 230 g (one A)
840 g − 230 g = 610 g (two Bs)
610 g ÷ 2 = 305 g (one B)
305 g + 230 g = 535 g
A and B weigh 535 g altogether.

Exercise 8 • pages 105–108

Exercise 8

Check

1. Fill in the blanks with m, cm, km, g, kg, L, or mL.

   (a) A hummingbird weighs about 4 __g__.

   (b) The lake is 100 __m__ long.

   (c) A dry bag for water sports has a capacity of 5 __L__.

   (d) A milk carton can hold 450 __mL__.

   (e) 10 train cars hooked together are about 1 __km__ long.

   (f) The wingspan of an eagle is about 180 __cm__.

   (g) An eagle weighs about 5 __kg__.

2. (a) 4 m 46 cm = __446__ cm

   (b) 4 km 46 m = __4,046__ m

   (c) 6 kg 207 g = __6,207__ g

   (d) 2 L 2 mL = __2,002__ mL

   (e) 4,689 g = __4__ kg __689__ g

   (f) 8,020 m = __8__ km __20__ m

   (g) 8,020 cm = __80__ m __20__ cm

3. 4 L 250 mL of water weighs __4,250__ grams.

4. 8 decks of cards weigh 768 g.
   How much does one deck of cards weigh?
   768 g ÷ 8 = 96 g
   One deck of cards weighs 96 g.

5. Javier Sotomayor has a high jump record of 2 m 45 cm.
   Stefka Kostadinova has a high jump record of 209 cm.
   Who has the greatest high jump record and by how much?
   245 cm − 209 cm = 36 cm.
   Javier Sotomayor has the greatest high jump record by 36 cm.

6. A cook had 6 bottles of olive oil, each of which contained 750 mL of olive oil.
   After two weeks, she had 3 L 450 mL of olive oil left.
   How much olive oil did she use in 2 weeks?
   Give your answer in compound units.
   6 × 750 mL = 4,500 mL
   4,500 mL − 3,450 mL = 1,050 mL
   She used 1 L 50 mL of olive oil in 2 weeks.

7. This diagram shows the distance between some places in a town.

   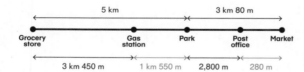

   (a) Which two places are separated by 7,800 m?
   5,000 m + 2,800 m = 7,800 m
   The grocery store and the post office.

   Students may convert to all meters, use mental math to add or subtract within a kilometer, or a combination of these two methods.

   (b) How much farther is it from the park to the market than from the park to the post office?
   3,080 m − 2,800 m = 280 m
   280 m

   (c) Owen went from the park to the gas station and then to the post office.
   How far did he travel?
   5 km − 3 km 450 m = 1 km 550 m
   1 km 550 m + 1 km 550 m + 2 km 800 m = 5 km 900 m
   5 km 900 m

   (d) What is the distance from the gas station to the market?
   5 km + 3 km 80 m = 8 km 80 m
   8 km 80 m − 3 km 450 m = 4 km 630 m
   4 km 630 m

   (e) Mila went from the gas station to 2 different places.
   She traveled a total of 7 km 150 m.
   Which 2 places did she go to and in what order?
   Use estimation to narrow down choices.
   5 km − 3 km 450 m = 1 km 550 m
   1 km 550 m + 2 km 800 m + 2 km 800 m = 7 km 150 m
   She went from the gas station to the post office to the park.

Challenge

8. 2 screwdrivers and 2 boxes of screws weigh 4 kg 632 g altogether.
   If 1 box of screws weighs 2 kg 267 g, how much does 1 screwdriver weigh?

   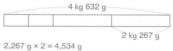

   2,267 g × 2 = 4,534 g
   4,632 g − 4,534 g = 98 g
   98 g ÷ 2 = 49 g
   1 screwdriver weighs 49 g.

9. Two containers have 950 mL of water altogether.
   Container B has 4 times as much water as container A.
   How much water has to be poured from container B into container A so they both have the same amount of water?

   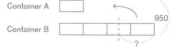

   Half the difference between the two containers needs to be transferred.
   950 mL ÷ 5 = 190 mL
   190 mL × 3 = 570 mL
   570 mL ÷ 2 = 285 mL
   285 mL of water.

10. One cup is 16 cm tall.
    Two stacked cups have a height of 20 cm.
    How many cups are needed to make a stack that is 80 cm high?

    The first cup is 16 cm and each additional cup is 4 cm.
    80 cm − 16 cm = 64 cm
    64 cm ÷ 4 cm = 16
    There are 16 units of 4 cm plus the first cup.
    17 cups are needed to make a stack that is 80 cm high.

Exercise 9 • pages 109–114

Exercise 9

Check

1
(a) How many tens make 9,400?
  940
(b) How many tenths make 2?
  20
(c) How many digits are there in the quotient of 581 ÷ 7?
  2
(d) Estimate the sum of 4,586 + 768.
  Answers may vary. Example: 4,600 + 800 = 5,400
(e) In the sum of 3,985 and 1,567, what digit is in the tens place?
  5
(f) For the difference between 42 and 6,003, what digit is in the hundreds place?
  9
(g) What is the remainder for 689 ÷ 2?
  1

2 Write the missing digits.

(a)
```
  6, 4 8 7
+ 2, 9 0 8
  9, 3 9 5
```

(b)
```
  6, 4 8 7
− 2, 9 0 8
  3, 5 7 9
```

(c)
```
    6 2 9
  ×     7
  4, 4 0 3
```

(d)
```
        8 7
    8)6 9 6
      6 4
        5 6
        5 6
         0
```

3 Write the numbers for all of the unlabeled tick marks shown on each number line.

(a) 0 50 100 150 200 250 300 350 400 450 **500** 550 600

(b) 0 $\frac{1}{7}$ $\frac{2}{7}$ $\frac{3}{7}$ $\frac{4}{7}$ $\frac{5}{7}$ $\frac{6}{7}$ 1 $\frac{8}{7}$ $\frac{9}{7}$ $\frac{10}{7}$

(c) 4,000 4,500 5,000 5,500 **6,000** 6,500 7,000

(d) 0 32 64 96 128 160 192 224 **256** 288 320 352 384
256 ÷ 8 = 32

4
(a) 8 × 8 = 16 + 16 + 32
(b) 482 = 96 × 5 + 2
(c) 620 × 9 = 620 × 10 − 620

5 Write the fractions in simplest form.

(a) $\frac{10}{16} = \frac{5}{8}$ (b) $\frac{6}{8} = \frac{3}{4}$

(c) $\frac{25}{50} = \frac{1}{2}$

6 The information below shows how far a paper airplane flew on 5 different trials.

Trial	1	2	3	4	5
Distance	1 m 80 cm	2 m 40 cm	90 cm	350 cm	260 cm

(a) Complete the bar graph with the information from the table.

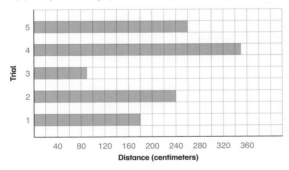

(b) Put the trials in order of least to greatest distance.
Trial 3, Trial 1, Trial 2, Trial 5, Trial 4

(c) What is the difference between the greatest and the least distance the plane flew?
350 cm − 90 cm = 260 cm
2 m 60 cm

7 The sum of three numbers is $\frac{7}{10}$.
The sum of the first and second numbers is $\frac{5}{10}$.
What is the third number?
$\frac{7}{10} - \frac{5}{10} = \frac{2}{10}$

$\frac{2}{10} = \frac{1}{5}$

The third number is $\frac{1}{5}$.

8 n is a number. Find the value of n.

$\frac{1}{n} = \frac{n}{16}$

n = 4

9 Is the product of 1 × 2 × 3 × 4 × .... × 19 × 20 even or odd?
It is even.
2 is a factor in this expression, so the total product will be even.

10 Sort the following fractions into 2 groups.
Explain why you grouped them that way.

$\frac{1}{5}$ $\frac{4}{5}$ $\frac{3}{9}$ $\frac{1}{3}$ $\frac{2}{6}$ $\frac{3}{5}$ Answers may vary. Possible answers provided.

Unit fractions: $\frac{1}{5}, \frac{1}{3}$; Non-unit fractions: $\frac{4}{5}, \frac{3}{9}, \frac{2}{6}, \frac{3}{5}$

Fractions in simplest form: $\frac{1}{5}, \frac{1}{3}, \frac{4}{5}, \frac{3}{5}$; Fractions not in simplest form: $\frac{3}{9}, \frac{2}{6}$

Fractions greater than one-half: $\frac{4}{5}, \frac{3}{5}$; Fractions less than one-half: $\frac{1}{5}, \frac{1}{3}, \frac{3}{9}, \frac{2}{6}$

© 2017 Singapore Math Inc.   Teacher's Guide 3B Chapter 11   137

11. Fuyu used $\frac{2}{3}$ m of string to tie a package, $\frac{2}{5}$ m for an art project, and $\frac{3}{10}$ m to tie up a tomato plant.
On which of the above items did she use the most string?
She used the most string tying the package.

12. (a) A men's shot put used in a track and field competition weighs 7 kg 260g.
Write its weight in grams.
7,260 g

(b) Randy Barnes set a world record of 23 m 12 cm for the shot put throw.
Write this distance in centimeters.
2,312 cm

13. The total capacity of 3 containers, A, B, and C, is 9 L 700 mL altogether.
The total capacity of container A and B is 5 L 800 mL altogether.
The capacity container C is 800 mL more than the capacity of container A.
What is the capacity of each of the containers in liters and milliliters?

Container C: 9,700 mL − 5,800 mL = 3 L 900 mL
Container A: 3,900 mL − 800 mL = 3,100 mL = 3 L 100 mL
Container B: 9,700 mL − 3,900 mL − 3,100 mL = 2,700 mL = 2 L 700 mL

Challenge

14. There are 2 different trails to a lookout.
The total distance up and back for both is 900 m altogether.
Trail A is 20 m longer one way to the lookout than Trail B.
How long is Trail B one way?
2 × 20 m = 40 m
900 m − 40 m = 860 m
860 m ÷ 4 = 215 m
Trail B is 215 m one way.
Check: 215 + 215 + 235 + 235 = 900

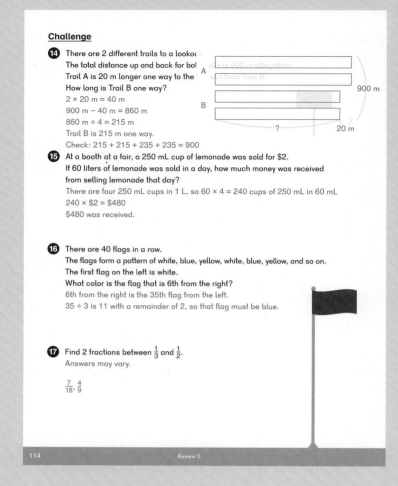

15. At a booth at a fair, a 250 mL cup of lemonade was sold for $2.
If 60 liters of lemonade was sold in a day, how much money was received from selling lemonade that day?
There are four 250 mL cups in 1 L, so 60 × 4 = 240 cups of 250 mL in 60 mL
240 × $2 = $480
$480 was received.

16. There are 40 flags in a row.
The flags form a pattern of white, blue, yellow, white, blue, yellow, and so on.
The first flag on the left is white.
What color is the flag that is 6th from the right?
6th from the right is the 35th flag from the left.
35 ÷ 3 is 11 with a remainder of 2, so that flag must be blue.

17. Find 2 fractions between $\frac{1}{3}$ and $\frac{1}{2}$.
Answers may vary.

$\frac{7}{18}, \frac{4}{9}$

Chapter 12 Geometry — Overview

Suggested number of class periods: 7–8

	Lesson	Page	Resources	Objectives
	Chapter Opener	p. 145	TB: p. 123	Investigate shapes.
1	Circles	p. 146	TB: p. 124 WB: p. 115	Identify the center, radius, and diameter of a circle and understand their relationships.
2	Angles	p. 149	TB: p. 128 WB: p. 119	Learn the parts of angles. Identify angles in shapes.
3	Right Angles	p. 151	TB: p. 132 WB: p. 121	Identify right angles. Compare the size of right angles to the size of other angles.
4	Triangles	p. 154	TB: p. 137 WB: p. 125	Explore the side properties of triangles. Classify triangles based on the lengths of their sides.
5	Properties of Triangles	p. 156	TB: p. 140 WB: p. 127	Explore the angle properties of triangles. Draw triangles with 1, 2, or no equal sides using set squares or circle dot paper.
6	Properties of Quadrilaterals	p. 158	TB: p. 143 WB: p. 130	Explore the side and angle properties of quadrilaterals. Identify quadrilaterals with right angles as rectangles.
7	Using a Compass	p. 161	TB: p. 147 WB: p. 134	Extension: Use a compass to draw shapes.
8	Practice	p. 164	TB: p. 151 WB: p. 139	Practice concepts from the chapter.
	Workbook Solutions	p. 167		

Chapter 12 Geometry

Notes

In Dimensions Math 2B, students classified triangles, quadrilaterals, pentagons, and hexagons by the number of sides. They learned that a polygon is a closed shape with straight sides and that the number of sides is equal to the number of corners. Students also partitioned a circle into quarter-circles and half-circles.

In this chapter, students begin with the geometry of circles, naming and measuring the center, radius, and diameter of a circle. Experience with circles is the foundation for understanding angles, because angles open around a circle.

Students will then learn to identify angles and right angles, and investigate the size of angles by comparing them to right angles. After studying angles, students will classify triangles and quadrilaterals by their angles and side lengths.

Circles

Lesson 1: Circles helps students find the center of a circle and understand that the edge of a circle is always the same distance from the center. Any line segment from the edge to the center is a radius. The term radius comes from Latin (radius) meaning "spoke" (of a wheel).

The diameter is a line segment from a point on the circle, through the center, to the other side. The term diameter comes from Greek meaning "measuring across" (dia = across, metros = measure).

The length of the diameter is always twice the length of the radius.

Angles

An angle is formed when two lines intersect. The size of an angle is measured by determining how much one side is turned about the vertex compared to the other side.

We use the term "sides" to define the boundaries of an angle and the "vertex" to define the point about which the sides intersect. Students will not measure angles in Dimensions Math 3B.

Students will work with angles less than 180 degrees. They may, however, notice that a reflex angle (an angle greater than 180 degrees) is also an angle. Teachers should use arcs or colors to show which angle they are referring to:

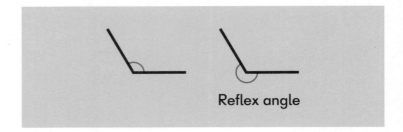

Reflex angle

In Lesson 2: Angles, students find and draw angles. Throughout the chapter, students use set squares, which are a pair of right-angled triangular rulers used for drawing:

Angles can be thought of as shapes or motions. When you model an angle with a manipulative, such as two cardboard strips connected at the vertex by a brad, you are modeling angles more as motions around a circle than as shapes.

Initially, students will see angles on the corners of polygons such as triangles.

Students will next classify three types of angles: right angles, angles greater than right angles, and angles less than right angles.

Chapter 12 Geometry

Notes

The terms "acute" and "obtuse" are not used in this chapter. The goal is to focus on the concepts. These terms will be taught in Dimensions Math 4.

Students should note that the size of an angle depends on the size of the opening, not the length of the sides.

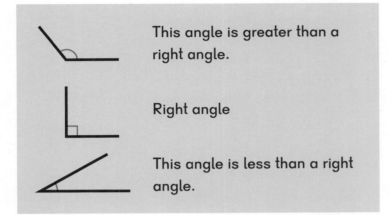

This angle is greater than a right angle.

Right angle

This angle is less than a right angle.

Note that hash marks are used to denote equal length sides or equal angles.

Triangles can also be classified by their angles:

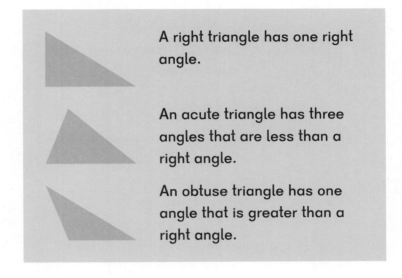

A right triangle has one right angle.

An acute triangle has three angles that are less than a right angle.

An obtuse triangle has one angle that is greater than a right angle.

Triangles

In Lesson 4: Triangles and Lesson 5: Properties of Triangles, students will investigate properties of triangles, classifying them by the size of their angles and the length of their sides.

Triangles can be classified by the lengths of their sides.

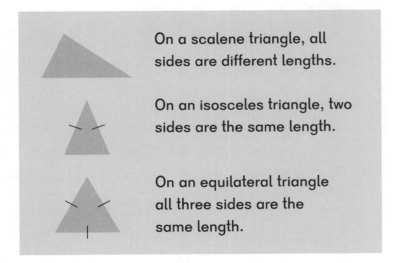

On a scalene triangle, all sides are different lengths.

On an isosceles triangle, two sides are the same length.

On an equilateral triangle all three sides are the same length.

The terms acute, obtuse, scalene, isosceles, and equilateral will be introduced in later grades.

Quadrilaterals

In Lesson 6: Properties of Quadrilaterals, students will extend their understanding of types of angles to quadrilaterals and informally explore the angles of quadrilaterals. Since students will not study parallel and perpendicular lines until Dimensions Math 4, parallelograms and trapezoids, or their properties, are not included in this chapter.

Chapter 12 Geometry

Notes

Students can now further their geometric understanding to see squares as a subcategory of rectangles because a square is a quadrilateral with 4 right angles and 4 sides equal in length.

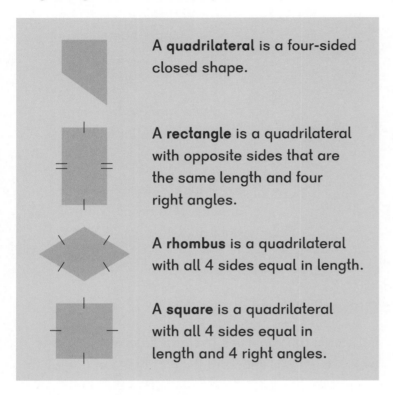

A **quadrilateral** is a four-sided closed shape.

A **rectangle** is a quadrilateral with opposite sides that are the same length and four right angles.

A **rhombus** is a quadrilateral with all 4 sides equal in length.

A **square** is a quadrilateral with all 4 sides equal in length and 4 right angles.

Note that a square is both a rectangle and a rhombus. All squares are rhombuses, but not all rhombuses are squares.

It is possible that students may draw a concave quadrilateral:

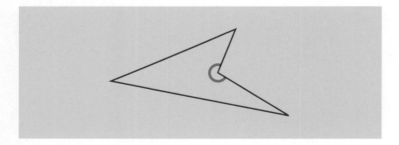

A quadrilateral is defined as a polygon with 4 sides and not by the lengths of sides. Students may draw quadrilaterals with concave sides, and should recognize that they are quadrilaterals, even though angles greater than 180 are not formally introduced at this level.

Note: Lesson 7: Using a Compass is an optional extension lesson that is included as a mathematical exploration of how to construct circles and triangles using a compass.

Drawing and using a compass for geometric shapes helps students focus on the attributes of a shape in greater detail and will be helpful when students learn formal constructions in later grades.

This lesson requires the use of a compass. A traditional compass will require a centimeter ruler to open the compass to the length of the radius.

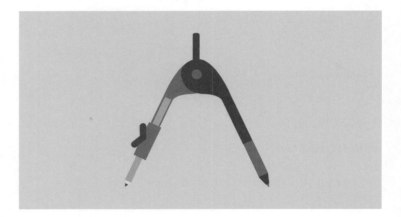

Students could also use a safety Bullseye compass:

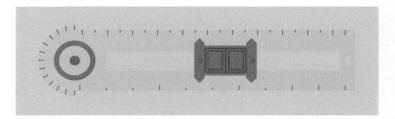

Students will need instruction and practice using a compass to draw a circle. When demonstrating, try to complete the circle smoothly without lifting the pencil from the paper.

Chapter 12 Geometry

Materials

- Brads or fasteners
- Cardstock, cardboard, or foam strips (or geostrips)
- Classroom rulers (inch and centimeter)
- Compass
- Dot paper
- Geoboards
- Half sheet of paper
- Pipe cleaners
- Rubber bands
- Set squares
- Straightedges
- Straws cut to the lengths 10 cm, 8 cm, 6 cm, and 5 cm
- Tangram sets
- Yardsticks or meter sticks
- Whiteboards

Blackline Masters

- Centimeter Graph Paper
- Centimeter Ruler
- Circle
- Circle Dot Paper
- Isometric Grid Paper
- Tangram
- Triangle

Storybooks

- *Grandfather Tang's Story* by Ann Tompert

Activities

Fewer games and activities are included in this chapter, as students will be drawing and classifying shapes and angles. The included activities can be used after students complete the **Do** questions, or anytime additional practice is needed.

Notes

Chapter Opener

Objective
- Investigate shapes.

Lesson Materials
- Tangram sets or Tangram (BLM)
- *Grandfather Tang's Story* by Ann Tompert

Provide students with a tangram set or Tangram (BLM) and have them make pictures.

This informal introduction to shapes will be revisited throughout the chapter when students identify angles, right angles, triangles, and quadrilaterals.

This Chapter Opener can be extended to a full lesson by further investigating tangrams or with a brief introductory activity before moving to Lesson 1: Circles.

If extending to a full lesson, *Grandfather Tang's Story* by Ann Tompert is a wonderful exploration of tangrams.

Lesson 1 Circles

Objective

- Identify the center, radius, and diameter of a circle and understand their relationships.

Lesson Materials

- Circle (BLM) (18-cm diameter)
- Centimeter Ruler (BLM) or classroom ruler

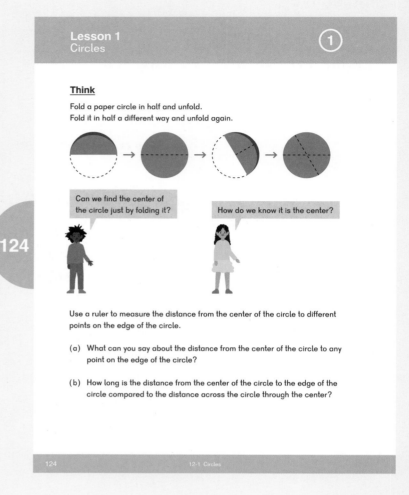

Think

Provide each student with a Circle (BLM) and a ruler. Ask students how they might find the center of the circle. Following the **Think** example, have students fold the circle in half, then unfold and fold in half again, a different way. Have them draw a line on each crease.

After folding the circle in half the second time, students will see the intersection of the two lines (or creases). Ask them to use a ruler to find the distance from the intersection to the edge of the circle in many different places.

By doing so, they will find out that all the distances from the intersection of the two lines to the edge of the circle is the same length. The intersection of two lines formed when the circle is folded in half is the center of the circle.

146   Teacher's Guide 3B Chapter 12   © 2017 Singapore Math Inc.

Learn

Have students label the center, diameter, and a radius on their circles.

Discuss the different terms. Students should compare the folds and lines on their paper circles to Alex, Emma, and Mei's comments.

All lines through the center of a circle are diameters and all lines from the center to the edge of a circle are radii.

Ensure students understand that the diameter is always twice the length of the radius of a circle. This can be shown by folding the circle in half and then opening it multiple times.

Students should note that for all of the folds on their circles, the length from the center to the edge of the circle, the radius, is 9 cm so all diameters are twice that length, 18 cm.

Do

1. Ensure students notice that if they find one measurement with a ruler, they can find the second measurement using calculation.

2. If students know the radius, they can multiply it by 2 to get the diameter. If they know the diameter, they can divide it by 2 to get the radius.

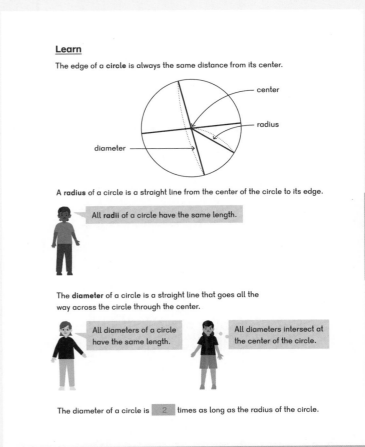

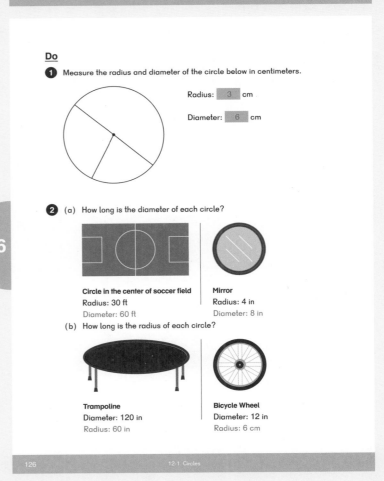

❸ Students should see that the sides of the square also show the diameter of the circle where the circle touches the square.

They can divide 12 inches by 2 to find the radius.

❹ Students can think about the circle they folded in **Think**. If they unfold the quarter-circle into a half-circle, they can find the diameter by multiplying the radius by 2.

❺ Help students see that the small circles have the same diameter, so the point where the small circles meet is the center of the larger circle. The diameter of one smaller circle is the radius of the larger circle.

If a radius of the smaller circle is 8 cm, students can find the diameter of that circle as 8 cm × 2, or 16 cm. The diameter of the larger circle then is 16 cm × 2, or 32 cm.

Students may also see that the diameter of the large circle is equal to 4 units of a radius of the small circles, or 4 × 8 cm.

Exercise 1 • page 115

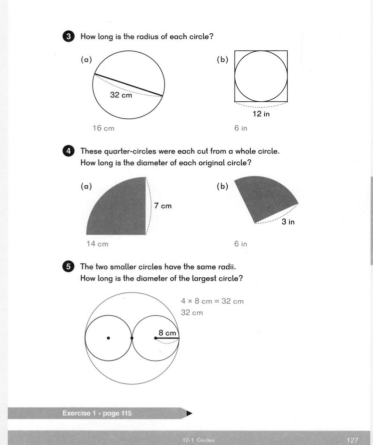

Lesson 2 Angles

Objectives
- Learn the parts of angles.
- Identify angles in shapes.

Lesson Materials
- Cardstock, cardboard, or foam strips (or geostrips)
- Brads or fasteners
- Set squares
- Geoboards
- Rubber bands

Think

Have students fasten together 2 strips of cardstock with a brad or fastener similar to the image in **Think**. Discuss the term "angle" and relate it to the term "corner" used in Dimensions Math 2B.

Have students find angles in shapes in the classroom. Try to find angles that are not just right angles. Ask students to share the angles they found.

Learn

Explain that an angle is formed when you open the sides, or turn one of them, about the vertex. Discuss the terms "vertex" and "side," as well as Mei, Sofia, and Dion's comments.

When the two sides of an angle form a straight line, it is called a "straight angle".

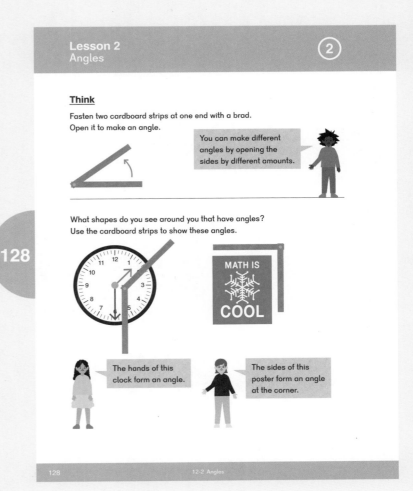

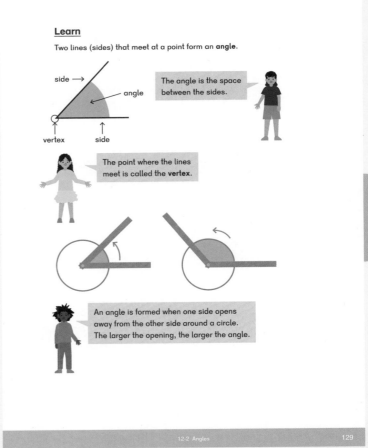

Do

3 – 5 To reduce the number of manipulatives required, groups of students could work on all three problems at once, taking turns with the set squares and geoboards.

After each student has had an opportunity to use each tool, the groups can discuss the different angles that students made.

Exercise 2 • page 119

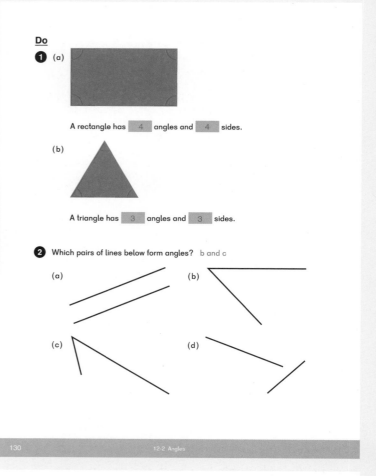

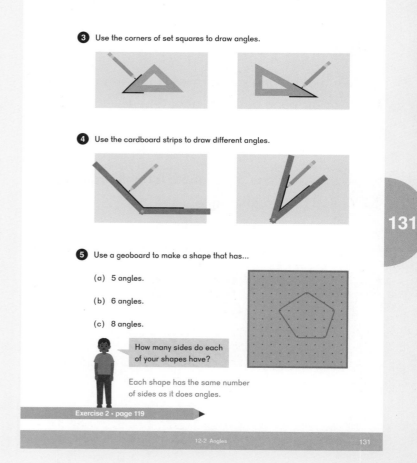

Lesson 3 Right Angles

Objectives

- Identify right angles.
- Compare the size of right angles to the size of other angles.

Lesson Materials

- Half sheet of paper
- Cardboard strip angle-maker from Lesson 2
- Set squares
- Tangram sets or Tangram (BLM) (optional)
- Yardsticks or meter sticks

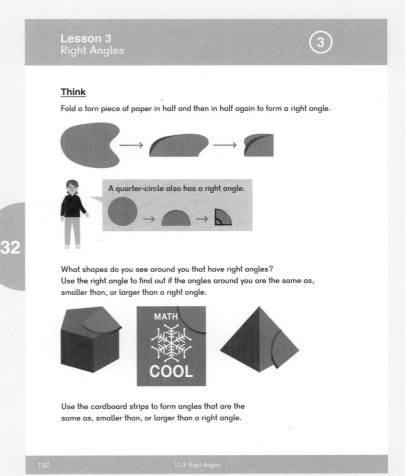

Think

Provide students with a half sheet of paper and have them tear off edges until they have a non-standard shape with rounded corners similar to what is shown in **Think**.

By folding the shape twice on top of itself, as shown in **Think**, they will create a right angle. Discuss Emma's comment.

Have students discuss the shapes they started with and then how they folded them so they can see, regardless of the starting shape, that they can always fold a right angle.

Have students use their right angle paper to find angles in the classroom that are equal to and greater or less than a right angle.

Discuss the different objects in the classroom that students measured. Have them show angles that are smaller than and larger than a right angle.

© 2017 Singapore Math Inc.   Teacher's Guide 3B Chapter 12   151

Learn

Discuss Dion's comment about the size of the angle. Remind students the angle is made by holding one side and turning the other side around the vertex. The amount of turn is the size of the angle.

If necessary, use two meter sticks or yardsticks to make a right angle. Compare that right angle to the cardboard strips set to a right angle. The longer sides created by the meter sticks do not mean that the angle is greater than the angle created by the shorter cardboard strips. The actual angle is the same.

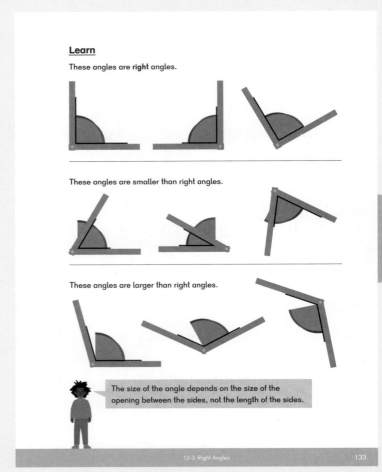

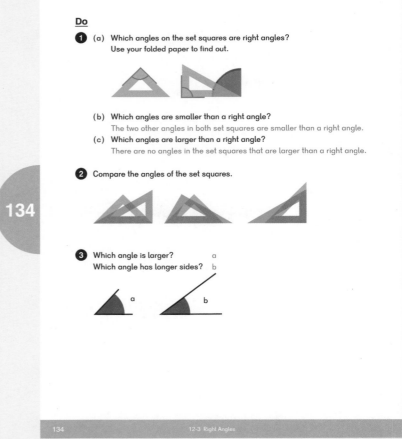

Do

4. To extend, have students put together tangram pieces so that the composed figure has a right angle.

5–7 Students can use their folded paper or the right angle on set square to measure the angles.

Activity

Provide students with tangrams. Challenge them to create:

- a square using two tangram pieces
- a non-square rectangle using four tangram pieces
- a figure shaped like a house using all seven tangram pieces

Exercise 3 • page 121

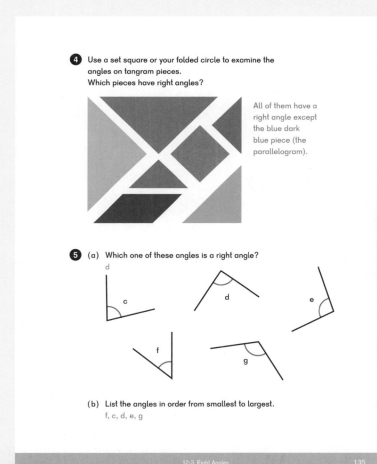

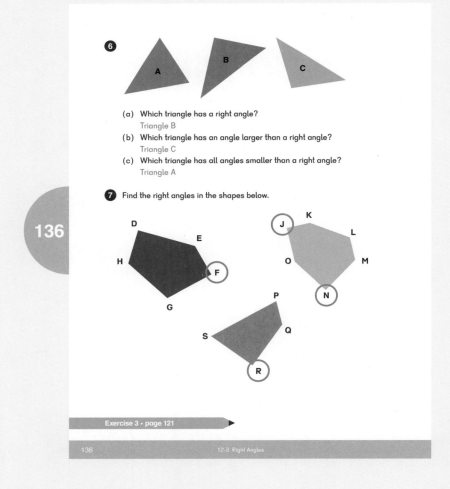

Lesson 4 Triangles

Objectives

- Explore the side properties of triangles.
- Classify triangles based on the lengths of their sides.

Lesson Materials

- Straws cut to the lengths 10 cm, 8 cm, 6 cm, and 5 cm
- Pipe cleaners
- Rulers
- Cardboard strip angle-maker from Lesson 2

Think

Provide students with straws as described in the **Materials** list and pipe cleaners. Have students make different triangles and try to group them according to the length of their sides.

The pipe cleaners can be used to connect the straws:

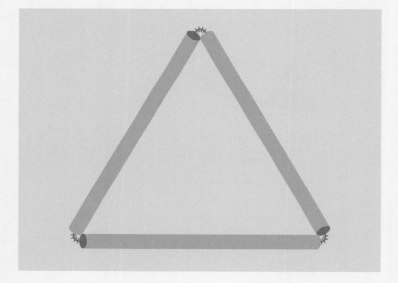

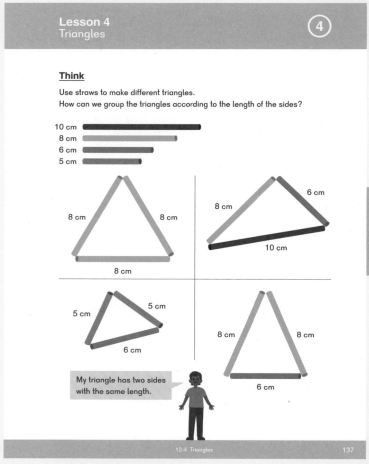

Learn

Have students sort their triangles according to the length of the sides as shown in **Learn**. They can share and discuss their different triangles, and how they are grouped.

Have students use one of their angle-makers (the cardboard strips made in Lesson 2) to see if they can construct triangles with angles larger than a right angle or smaller than a right angle.

Discuss Emma's comment and have students prove her statement by trying to build a triangle with the straws. Ask them if they can create a triangle with two angles that are larger than a right angle and have them explain their answers.

The sum of any two sides must have a total length longer than the third side.

Do

③ Discuss Sofia's comment on the tick marks denoting sides of equal length. To understand the single or double marks, draw a rectangle to show that the same number of marks are used to denote equal length sides. For example:

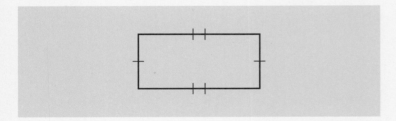

Exercise 4 • page 125

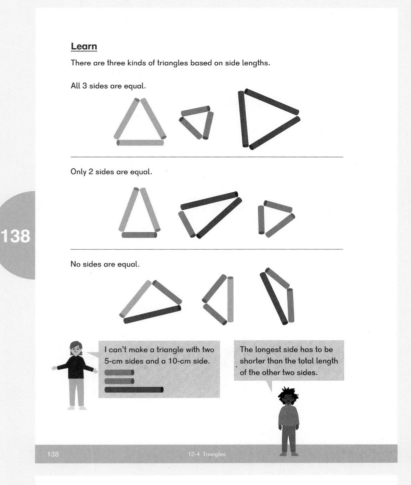

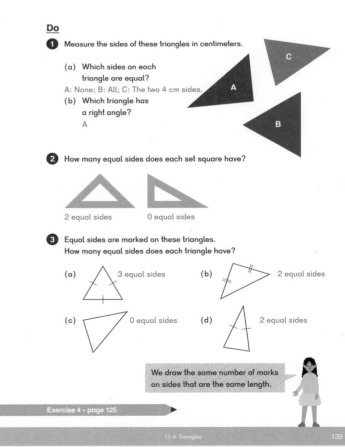

Lesson 5 Properties of Triangles

Objectives

- Explore the angle properties of triangles.
- Draw triangles with 1, 2, or no equal sides using set squares or circle dot paper.

Lesson Materials

- Triangle (BLM)
- Set squares
- Circle Dot Paper (BLM)
- Straightedge or ruler

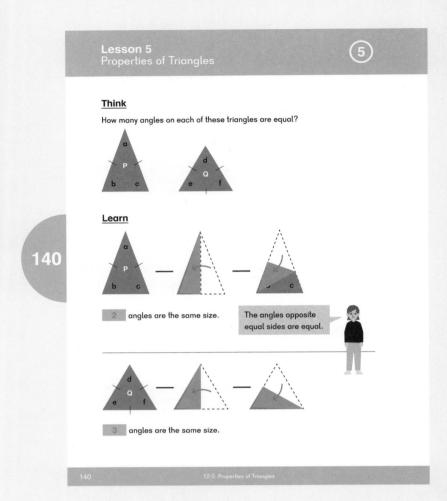

Think

Provide each student with a Triangle (BLM) and have them cut out the triangles. Make sure that students notice that each angle is labeled with a different letter.

Pose the **Think** problem and have students fold or cut the triangles to find which angles on each triangle are equal.

Learn

Students should note that on the first triangle, two sides are the same length and two angles are the same size.

In the second triangle, all three sides are the same length and all three angles are the same size.

Discuss Emma's comment. Help students understand that if two sides of a triangle are equal then two angles are equal. The reverse is also true. If two angles are equal, then the two sides adjacent to the angles are equal. This is a property of isosceles triangles. (Students do not need to know the term "isosceles".)

Do

① Students can work with partners or take turns using the set squares.

②—③ Provide students with Circle Dot Paper (BLM) and a straightedge.

② (a) If all of the sides are the same length, all of the angles will be the same.

③ If you draw a triangle using the center of the circle as a vertex and the other two points of the triangle are on the circle, then at least two of the sides will be the same length because they are radii.

Exercise 5 • page 127

Do

① Put two set squares together to make triangles. Trace around them.

The three triangles shown are examples. Answers will vary depending on the triangles students make.

(a) How many different triangles can you make?
(b) Which sides of each triangle have the same length?
(c) Which angles of each triangle are the same size?
(d) Which triangle is a right triangle with two equal sides?

② Use circle dot paper to draw triangles by connecting the dots on the edge of the circles.

(a) Draw some triangles with three equal sides.
Which angles are equal?   All 3 of the angles are equal.

(b) Draw some triangles with two equal sides.
Which angles are equal?   The two angles opposite the equal sides are equal.

Triangles shown on this page are examples.

Do any of them have a right angle?

(c) Draw some triangles with no equal sides.

Do any of them have a right angle?

If students used a diameter as one of the sides in (b) or (c), their triangle will have a right angle.

③ Draw triangles using the center of the circle for one vertex. What is the same about all of them?

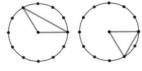

Two of the sides are radii...

They all have at least 2 equal sides and angles.

Exercise 5 • page 127

Lesson 6 Properties of Quadrilaterals

Objectives

- Explore the side and angle properties of quadrilaterals.
- Identify quadrilaterals with right angles as rectangles.

Lesson Materials

- Geoboard
- Rubber bands
- Tangram set or Tangram (BLM)
- Set squares or folded paper angle-maker from Lesson 3
- Dot paper
- Centimeter ruler

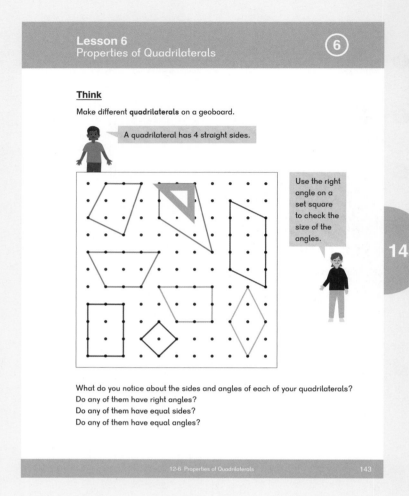

Think

Provide students with geoboards, rubber bands, and dot paper. Have students work on the **Think** task and discuss Alex's definition of a quadrilateral.

Have students record the quadrilaterals they make on the geoboard onto the dot paper. They can use a set square or their folded paper right angle checker from Lesson 3 to see which ones have right angles.

They can use a centimeter ruler to check the length of the sides if needed.

Learn

Discuss the examples with students. Quadrilaterals are closed shapes with 4 straight sides.

Have students note the tick marks denoting sides of the same length on the rectangle and square. Discuss the relationship between a rectangle, a square, and a rhombus as shared by Dion, Sofia, and Mei.

Since a rectangle is defined as a quadrilateral with 4 right angles, a square fits the definition as well. A square is a rectangle with 4 equal sides.

Remind students that all squares are rectangles, and some rectangles are squares.

Mei introduces the term "rhombus" for a shape with 4 sides of equal length.

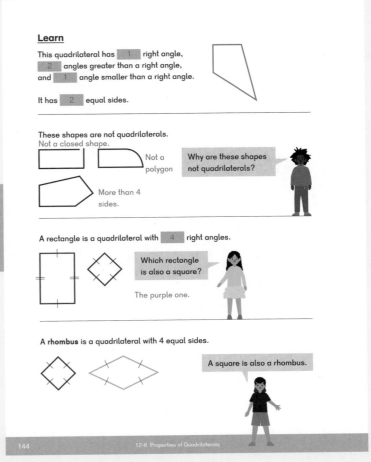

Do

1 Students can use more than 2 tangram pieces to make quadrilaterals. Have students share the different quadrilaterals they make.

2 Students will find other examples not included in the textbook.

3 Students can use a centimeter ruler or a strip of paper to compare the lengths of the sides.

Exercise 6 • page 130

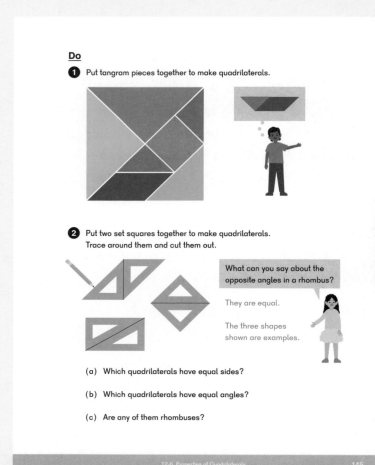

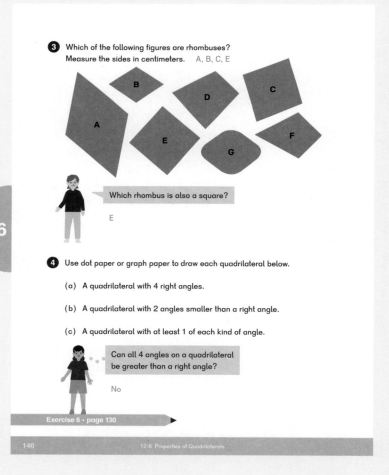

Lesson 7 Using a Compass

Objective

- Extension: Use a compass to draw shapes.

Lesson Materials

- Compass
- Centimeter ruler
- Centimeter Graph Paper (BLM)

Think

Introduce a compass to students and explain that it is a tool for drawing circles and comparing lengths. Allow students some time to experiment using a compass to draw circles.

Demonstrate how to use the compass to draw a circle with a given radius:

- Open the compass to a radius of 5 cm. Check the width by using a ruler.

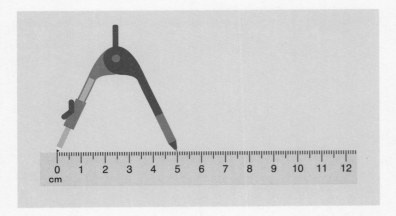

- Place the needle where the center of the circle should go.
- Twist your wrist to draw a circle. Try to make a complete circle without lifting the pencil.
- Provide students with a compass and Centimeter Graph Paper (BLM) and have them practice using the compass to draw the 5 cm radius and 8 cm diameter circles. Dion prompts students to find a radius when given the diameter.

Students may find **Think** (c) and Alex's thought challenging.

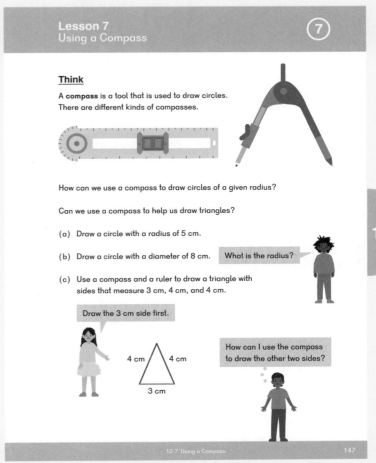

© 2017 Singapore Math Inc.   Teacher's Guide 3B Chapter 12   161

Learn

To help students draw a line that is 3 cm long, provide them with Centimeter Graph Paper (BLM).

For part (c), have students place the compass on each endpoint of the 3-cm base and draw an entire circle around each point. There will be two intersections points, one above the 3-cm line and one below.

The point, or vertex, where these arcs intersect is the top or bottom of the triangle. Students can connect the dots to make the triangle.

Because this point is on the edge of both circles of radius 4 cm, then the sides of the triangle are of length 4 cm.

Learn

(a) Make sure the distance from the center to the pencil point is 5 cm. Then turn the compass to draw the circle.

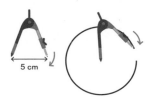

(b) The distance from the center to the pencil point should be 4 cm to draw a circle with a diameter of 8 cm.

(c) Draw the 3 cm side. Then use the compass to find out where to put the third vertex.

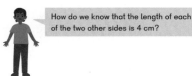

How do we know that the length of each of the two other sides is 4 cm?

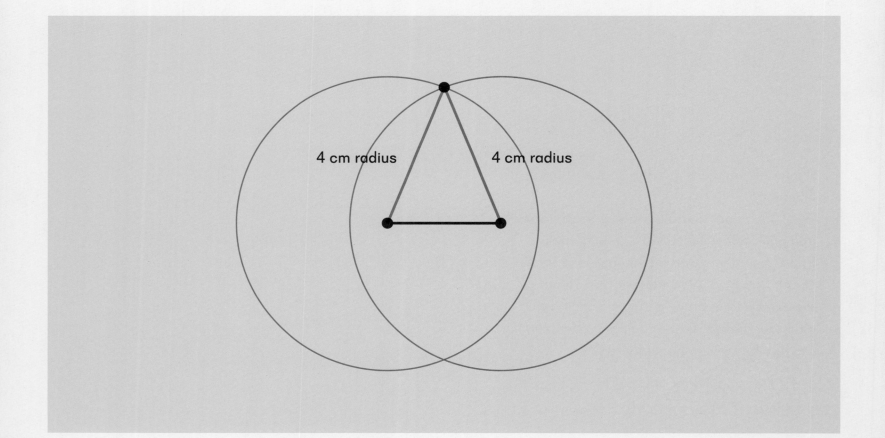

Do

Centimeter graph paper will make it easy for students to draw the lengths necessary for each problem.

5 Note that to produce the same designs that Mei has, students can think about where on the graph paper the centers of the circles or partial circles (arcs) need to be located.

This is a more challenging task for students than just creating their own interesting designs.

Exercise 7 • page 134

Lesson 8 Practice

Objective

- Practice concepts from the chapter.

Lesson Materials

- Set squares or angle-makers from Lesson 3
- Circle Dot Paper (BLM)

After students complete the **Practice** in the textbook, have them continue to practice by identifying angles and shapes from this chapter in the world around them.

Activity

▲ **Shape Draw**

Materials: Isometric Grid Paper (BLM)

Use Isometric Grid Paper (BLM) to draw the following polygons:

- Rectangle
- Rhombus
- Quadrilateral that is not a rectangle
- Triangle with a right angle
- Triangle with two sides of the same length
- Triangle with no sides of the same length
- Pentagon
- Hexagon
- Octagon

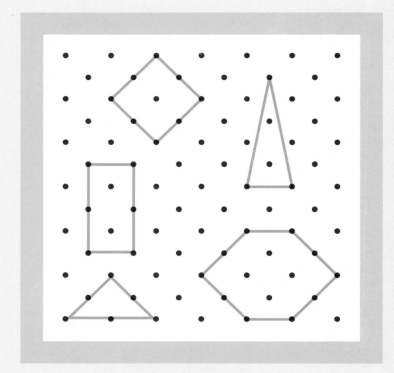

164   Teacher's Guide 3B Chapter 12   © 2017 Singapore Math Inc.

④—⑤ Have students use a folded paper from Lesson 3 or a set square to check the angles on the shapes.

Exercise 8 • page 139

Brain Works

★ Triangles

Materials: Geoboards, rubber bands

Two players take turns placing a rubber band around 3 pegs to make a triangle on the geoboard. A rubber band can share a peg with other rubber bands, but the triangles must not overlap (except along the edges and around the pegs).

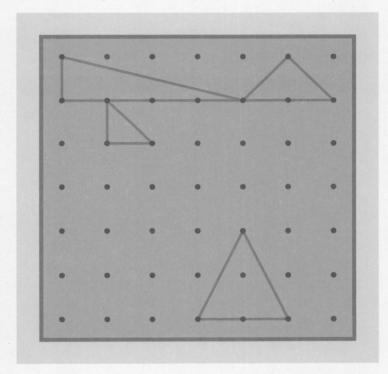

A player loses when he cannot make a triangle on his turn.

Extend by making quadrilaterals on the geoboard. How does the winning strategy change?

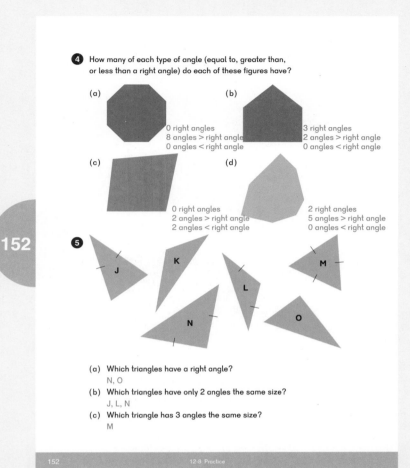

© 2017 Singapore Math Inc.　　Teacher's Guide 3B Chapter 12　　165

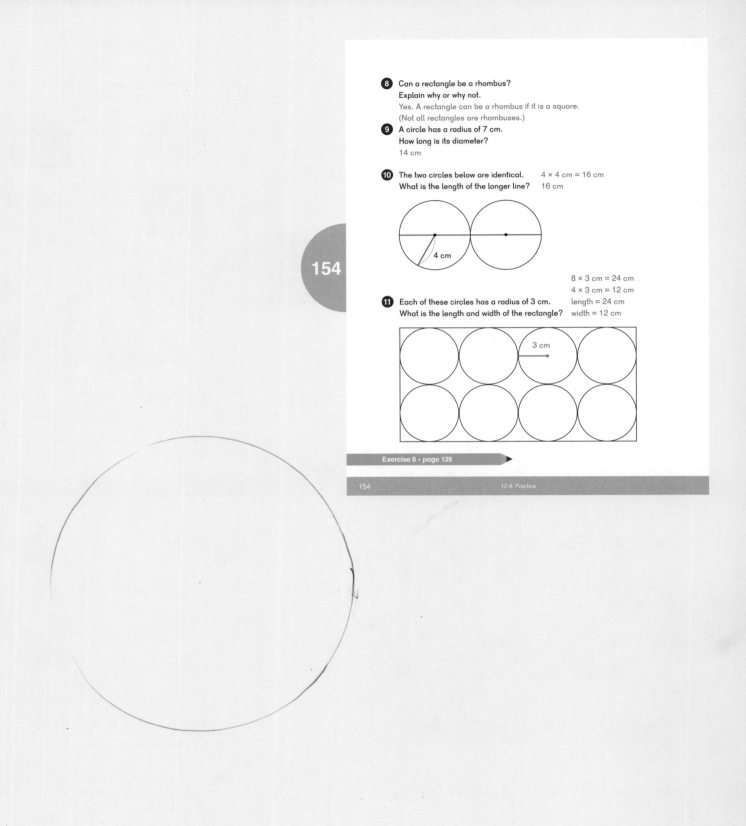

Exercise 1 • pages 115–118

Chapter 12 Geometry

Exercise 1

Basics

1. (a) Trace the lines that are diameters. (b) Trace the lines that are radii.

2. Measure the radius and diameter of each circle in centimeters.

   (a) Radius: 5 cm   Diameter: 10 cm

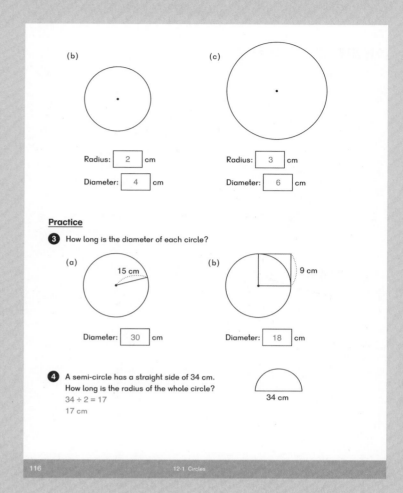

(b) Radius: 2 cm   Diameter: 4 cm

(c) Radius: 3 cm   Diameter: 6 cm

Practice

3. How long is the diameter of each circle?

   (a) 15 cm   Diameter: 30 cm

   (b) 9 cm   Diameter: 18 cm

4. A semi-circle has a straight side of 34 cm. How long is the radius of the whole circle?
   34 ÷ 2 = 17
   17 cm

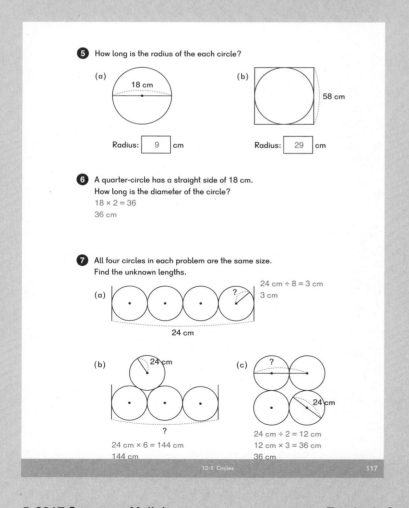

5. How long is the radius of the each circle?
   (a) 18 cm   Radius: 9 cm
   (b) 58 cm   Radius: 29 cm

6. A quarter-circle has a straight side of 18 cm. How long is the diameter of the circle?
   18 × 2 = 36
   36 cm

7. All four circles in each problem are the same size. Find the unknown lengths.
   (a) 24 cm   24 cm ÷ 8 = 3 cm   3 cm
   (b) 24 cm   24 cm × 6 = 144 cm   144 cm
   (c) 24 cm   24 cm ÷ 2 = 12 cm   12 cm × 3 = 36 cm   36 cm

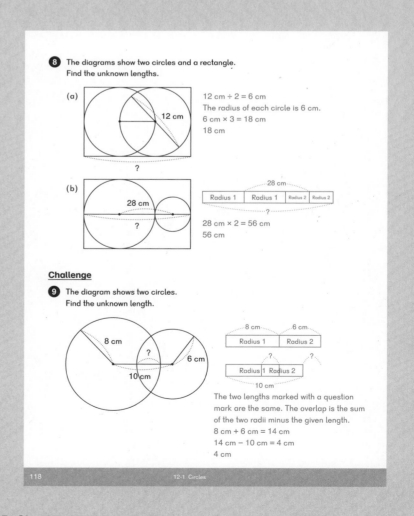

8. The diagrams show two circles and a rectangle. Find the unknown lengths.

   (a) 12 cm
   12 cm ÷ 2 = 6 cm
   The radius of each circle is 6 cm.
   6 cm × 3 = 18 cm
   18 cm

   (b) 28 cm
   28 cm × 2 = 56 cm
   56 cm

Challenge

9. The diagram shows two circles. Find the unknown length.
   8 cm, 6 cm, 10 cm
   The two lengths marked with a question mark are the same. The overlap is the sum of the two radii minus the given length.
   8 cm + 6 cm = 14 cm
   14 cm − 10 cm = 4 cm
   4 cm

Exercise 2 • pages 119–120

Exercise 2

Basics

1. The diagram shows the angles formed between two strips of cardboard as one strip is opened away from the other.

    (a) Angle ___z___ is the largest.

    (b) The vertex of each angle is at the ___center___ of the circle.

    (c) The two lines are the ___sides___ of the angle.

2. Check ✓ the pairs of lines that form angles.

    [✓ on second and third boxes]

3. Check ✓ the figures that have at least one angle.

    [✓ on first and third boxes]

Practice

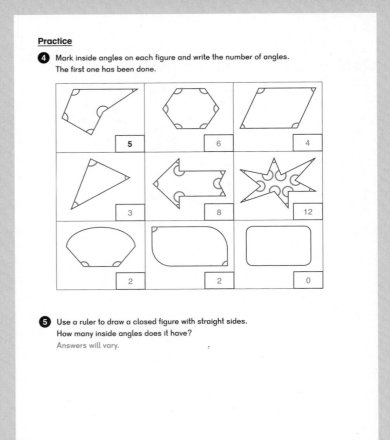

4. Mark inside angles on each figure and write the number of angles. The first one has been done.

    5, 6, 4
    3, 8, 12
    2, 2, 0

5. Use a ruler to draw a closed figure with straight sides. How many inside angles does it have?

    Answers will vary.

Exercise 3 • pages 121–124

Exercise 3

Basics

For any problem, you can use a set square or the corner of a rectangular card to compare the angles to a right angle.

① All rectangles have __4__ right angles.
How many right angles do each of the two triangles have?

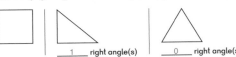

__1__ right angle(s)   __0__ right angle(s)

② The diagram shows the angles formed between two strips of cardboard as one strip is opened away from the other.

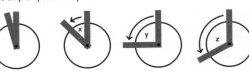

(a) Which of the labeled angles is smaller than a right angle?
x
(b) Which angle is a right angle?
y
(c) Which angle is larger than a right angle?
z
(d) Compare the angle on the right to the right angle above. Is it larger, smaller, or the same size as a right angle?
same size

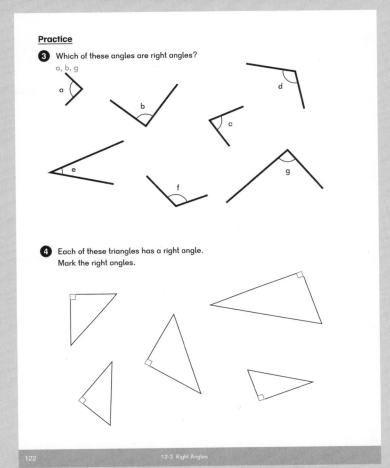

Practice

③ Which of these angles are right angles?
a, b, g

④ Each of these triangles has a right angle. Mark the right angles.

⑤
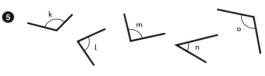

(a) Which angles are larger than a right angle?
k, o
(b) Which angles are smaller than a right angle?
l, n
(c) List the angles in order from smallest to largest.
n, l, m, k, o

⑥ Write how many right angles each figure has.
The right angles can be on the inside or on the outside of the figure.

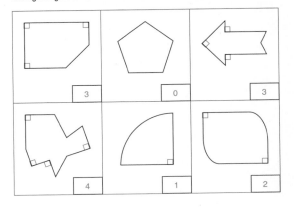

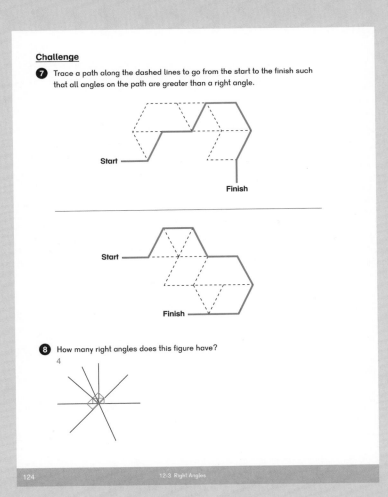

Challenge

⑦ Trace a path along the dashed lines to go from the start to the finish such that all angles on the path are greater than a right angle.

⑧ How many right angles does this figure have?
4

Teacher's Guide 3B Chapter 12

Exercise 4 • pages 125–126

Exercise 4

Basics

1. Measure the sides of the triangles in centimeters.

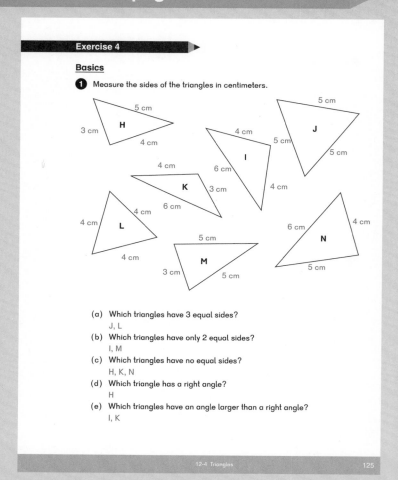

(a) Which triangles have 3 equal sides?
   J, L
(b) Which triangles have only 2 equal sides?
   I, M
(c) Which triangles have no equal sides?
   H, K, N
(d) Which triangle has a right angle?
   H
(e) Which triangles have an angle larger than a right angle?
   I, K

Practice

2. In each figure, sides with the same number of hash marks are equal in length, and all the circles have the same length radii.
Sort the labeled triangles according to the number of equal sides.

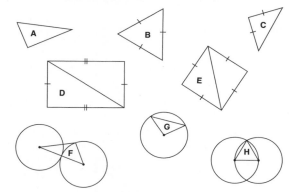

3 equal sides	Exactly 2 equal sides	0 equal sides
B, H	C, E, G	A, D, F

Challenge

The third side has to be shorter than the sum of the other two sides.

3. Is it possible to draw a triangle that has sides of the following measurements? Write "yes" or "no" next to each set of measurements.

(a) 5 cm, 4 cm, 11 cm
   no
(b) 2 cm, 6 cm, 5 cm
   yes
(c) 11 cm, 6 cm, 7 cm
   yes
(d) 8 cm, 3 cm, 5 cm
   no

Exercise 5 • pages 127–129

Exercise 5

Basics

1. Complete the sentences with 0, 1, 2, or 3.

(a) A triangle can have __0__, __2__, or __3__ equal angles.

(b) A triangle that has 2 equal sides has __2__ equal angles.

(c) A triangle that has 0 equal sides has __0__ equal angles.

(d) A triangle that has 3 equal sides has __3__ equal angles.

(e) A triangle with 1 right angle can have __0__ or __2__ equal sides.

Practice

2. Measure the sides of each triangle in centimeters and write the number of equal angles.

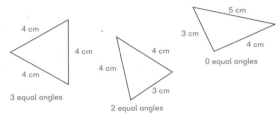

3. Use the circles and dots to help you draw three different triangles of each type.
All the circles are the same size so radii can be used to have sides of the same length.
Label your triangles, and then complete the table below.
Answers will vary.

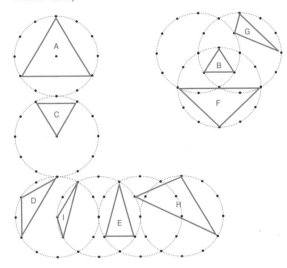

3 equal sides	Exactly 2 equal sides	0 equal sides
A, B, C	D, E, F	G, H, I

Challenge

4. The drawing shows how two toothpicks in a triangle made from six toothpicks can be moved to change one triangle into two smaller triangles.

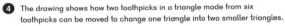

For each problem, all triangles should have three equal sides.

(a) Show how five toothpicks can be arranged to show two triangles.

(b) Show how three toothpicks can be removed to leave three triangles.

(c) Show how four toothpicks can be moved to show three triangles. The triangles do not have to be the same size.

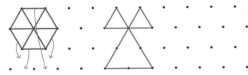

Exercise 6 • pages 130–133

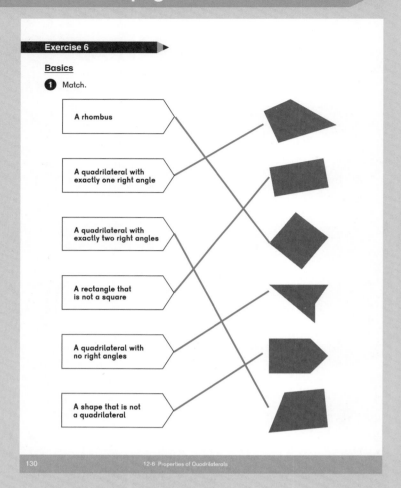

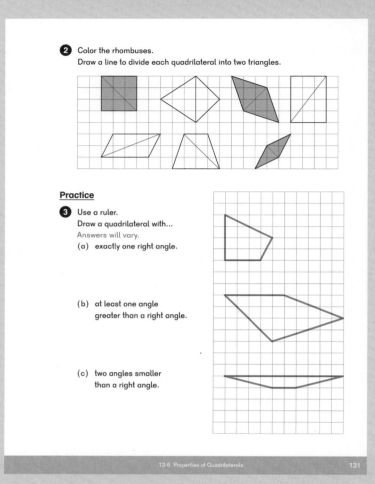

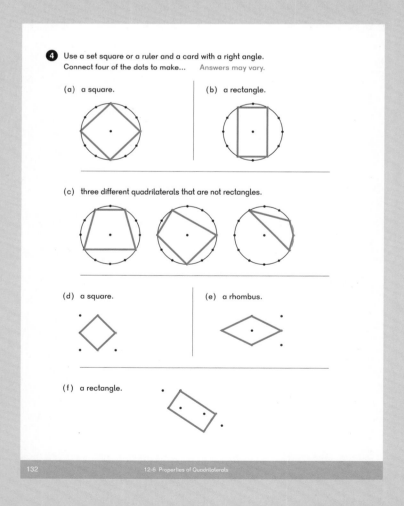

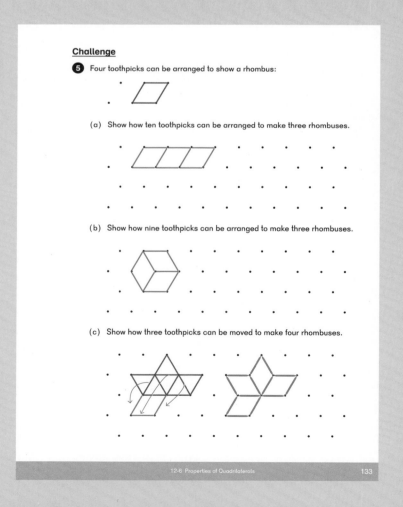

Exercise 7 • pages 134–138

Exercise 7

Basics

1. Use a compass and the centimeter graph below to draw a circle with a radius of...

   (a) 3 cm.

   (b) 5 cm.

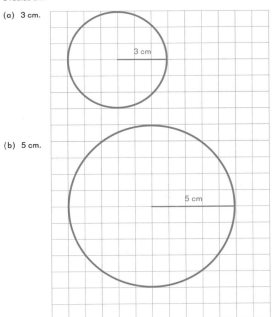

2. One side of a triangle and the radii of circles with centers at each end of that side are given.
Draw the other two sides of each triangle with lengths of...

   (a) 4 cm and 4 cm.

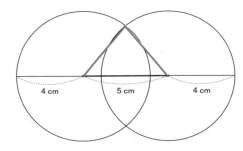

   (b) 3 cm and 2 cm.

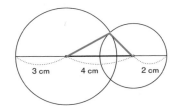

3. The diagram below shows a line that is 4 cm long.
Parts of two circles with centers at the ends of the line and radii of 4 cm are shown.
Draw the other two sides of a triangle to create a triangle with 3 equal angles.

Practice

4. Use a compass and the centimeter graph below to draw a triangle with sides equal to...

   (a) 9 cm, 6 cm, and 5 cm.

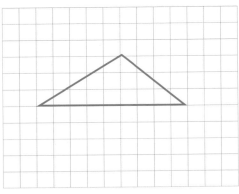

   (b) 5 cm, 6 cm, and 6 cm.

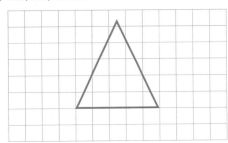

5. Draw a rhombus with sides of 5 cm, and a distance from one vertex to the opposite vertex of 3 cm.

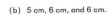

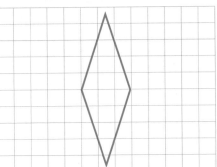

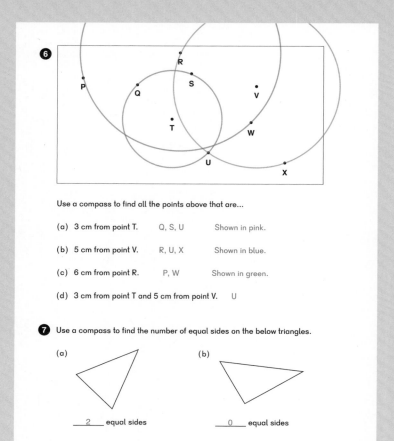

❻ Use a compass to find all the points above that are...

(a) 3 cm from point T.   Q, S, U   Shown in pink.

(b) 5 cm from point V.   R, U, X   Shown in blue.

(c) 6 cm from point R.   P, W   Shown in green.

(d) 3 cm from point T and 5 cm from point V.   U

❼ Use a compass to find the number of equal sides on the below triangles.

(a) __2__ equal sides

(b) __0__ equal sides

Exercise 8 • pages 139–142

Exercise 8

Check

1. A round satellite dish has a diameter of 28 cm.
   What is its radius?
   28 cm ÷ 2 = 14 cm
   14 cm

2. The rim of a basketball hoop has a diameter of 46 cm.
   What is its radius?
   46 cm ÷ 2 = 23 cm
   23 cm

3. Trace the two lines that form a right angle.

   (a)   (b)

4. Write "true" or "false."

   (a) All squares are rectangles.
       True
   (b) All rhombuses are squares.
       False; Example: A rhombus that does not have right angles.
   (c) All quadrilaterals can be cut into two triangles.
       True
   (d) All triangles have at least one angle less than a right angle.
       True
   (e) All quadrilaterals have at least one angle less than a right angle.
       False; Example: A rectangle.

5. (a) Complete the table below with the number of sides, angles, and types of angles.
   Only consider inside angles.

   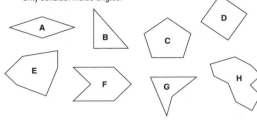

	Sides	Angles	Angles smaller than a right angle	Right angles	Angles larger than a right angle
A	4	4	2	0	2
B	3	3	2	1	0
C	5	5	0	0	5
D	4	4	0	4	0
E	6	6	1	2	3
F	6	6	2	1	3
G	4	4	3	0	1
H	10	10	0	3	7

   (b) List the shapes above that are quadrilaterals.
       A, D, G
   (c) List the shapes that are rhombuses.
       A, D

6. Each diagram shows two circles of the same size and one smaller circle. The centers of the circles and their intersections are marked with dots. Connect the dots to form…
   Answers may vary. Examples provided.

   (a) a triangle with 3 equal angles.
   (b) a triangle with 2 equal angles.
   (c) a triangle with 0 equal angles.
   (d) a rhombus.

   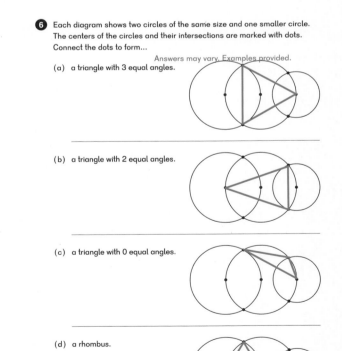

Challenge

7. How many rectangles can be made by joining the dots?

   10 rectangles can be made.
   4 ☐, 3 ☐, 2 ☐, and 1 ☐

8. How many rhombuses can be made by joining the dots?

   6 rectangles can be made.

9. Show how two toothpicks can be removed to make two squares.

10. Show how two toothpicks can be moved to make four squares of the same size.

© 2017 Singapore Math Inc.   Teacher's Guide 3B Chapter 12

Notes

Chapter 13 Area and Perimeter — Overview

Suggested number of class periods: 9–10

	Lesson	Page	Resources	Objectives
	Chapter Opener	p. 181	TB: p. 155	Investigate area.
1	Area	p. 182	TB: p. 156 WB: p. 143	Find the area of a figure in square units.
2	Units of Area	p. 184	TB: p. 160 WB: p. 146	Understand formal square units. Find the area of figures in formal square units.
3	Area of Rectangles	p. 186	TB: p. 163 WB: p. 150	Use multiplication to find the areas of rectangles and squares.
4	Area of Composite Figures	p. 188	TB: p. 166 WB: p. 153	Find the area of composite shapes formed of rectangles.
5	Practice A	p. 191	TB: p. 170 WB: p. 157	Practice finding the area of shapes.
6	Perimeter	p. 192	TB: p. 172 WB: p. 161	Understand the meaning of perimeter. Find the perimeter of polygons.
7	Perimeter of Rectangles	p. 194	TB: p. 175 WB: p. 164	Calculate the perimeter of rectangles given the length and width.
8	Area and Perimeter	p. 195	TB: p. 177 WB: p. 167	Understand the relationship between area and perimeter.
9	Practice B	p. 197	TB: p. 180 WB: p. 170	Practice finding area and perimeter.
	Workbook Solutions	p. 199		

Chapter 13 Area and Perimeter

Notes

In this chapter, students will be introduced to the concept of area. Students worked with standard units of length in Dimensions Math 2A Chapter 4: Length and Dimensions Math 3B Chapter 11: Measurement.

The student definition of area given in Lesson 1 is, "the size of a surface." Area is the space that a two-dimensional closed shape occupies.

Students begin by counting areas of figures made with square units.

In Lesson 2, students begin to find area in units of measurement, such as cm^2, in^2, m^2, ft^2, and yd^2. These units are read as "square centimeters," "square inches," etc.

Students will find the area of rectangles, with lengths and widths given in a whole number of standard units, by filling the area with square units. They will see that they can either multiply the number of squares along the length by the number of squares along the width, or vice versa.

Students will then derive the area of a rectangle as length × width or width × length, where the term "length" refers to the measurement of the longer side of the rectangle and the term "width" refers to the measurement of the shorter side. Students should realize from their informal experience with the commutative property that the order of the factors does not matter.

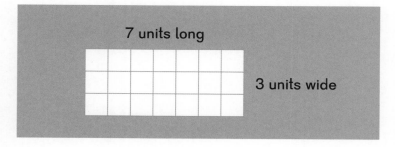

In Lessons 3 and 4, students will learn different strategies for finding the area of rectilinear composite figures.

Students will decompose composite figures to find the area of a shape made up of smaller rectangles.

For example:

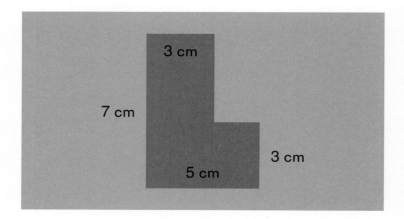

There are three strategies that can be used to find the area of this shape.

Strategy 1: Partitioning the rectangles into an upper and lower rectangle, students find:

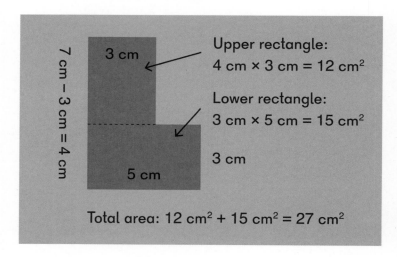

Upper rectangle:
4 cm × 3 cm = 12 cm^2

Lower rectangle:
3 cm × 5 cm = 15 cm^2

Total area: 12 cm^2 + 15 cm^2 = 27 cm^2

Strategy 2: Partitioning the rectangles into left and right, students find:

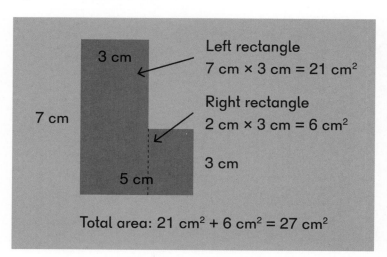

Left rectangle
7 cm × 3 cm = 21 cm^2

Right rectangle
2 cm × 3 cm = 6 cm^2

Total area: 21 cm^2 + 6 cm^2 = 27 cm^2

Chapter 13 Area and Perimeter

Strategy 3: Find the total area of the rectangle and subtract the missing part:

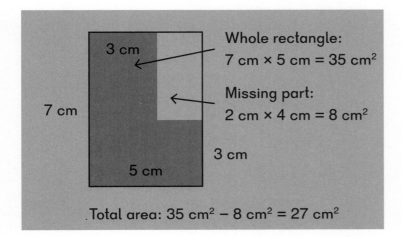

Whole rectangle:
7 cm × 5 cm = 35 cm²

Missing part:
2 cm × 4 cm = 8 cm²

Total area: 35 cm² − 8 cm² = 27 cm²

Students should understand that although the strategies look different, the underlying idea for finding the area of the composite figure is finding a way to distinguish the different rectangles in the shape. From there, students can use the area formula for a rectangle.

Areas can be added or subtracted as long as they are expressed with the same unit. This is similar to other measurements such as length, liquid volume, weight, time, money, etc.

Perimeter

Lesson 6 introduces the perimeter of a figure as the distance around the figure. Students will not be using any formulas to derive perimeter. When they can determine the length of the sides of a shape, they can determine the perimeter by adding the lengths of the sides together.

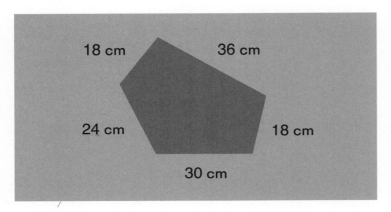

Students will learn different strategies to find the perimeter of rectangles. For example:

Strategy 1: Since opposite sides of a rectangle are equal, we know the lengths of all four sides and can add them:

7 cm + 3 cm + 7 cm + 3 cm = 20 cm

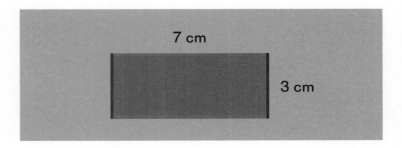

Strategy 2: Since opposite sides on a rectangle are equal, any rectangle has two sides of a given length and two sides of a given width. We can think of this as 7 + 7 + 3 + 3, or:

7 cm × 2 = 14 cm
3 cm × 2 = 6 cm
14 cm + 6 cm = 20 cm

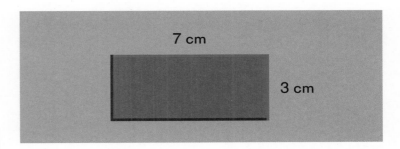

We can also add the length from one side and the width from one side together first, then multiply the sum by 2: (7 + 3) × 2, or:

7 cm + 3 cm = 10 cm
10 cm × 2 = 20 cm

Chapter 13 Area and Perimeter

Materials

- 1-in square tiles or pieces of paper
- 10-sided dice
- 4" × 6" index cards
- 6-sided dice
- Centimeter ruler
- Dry erase markers in 2 colors
- Dry erase sleeve
- Graph paper
- Rectangular towels or blankets of different but similar sizes
- Scissors
- Toothpicks or craft sticks
- Whiteboards

Blackline Masters

- Area of Composite Figures
- Centimeter Graph Paper
- Inch Graph Paper
- Square and Rectangle

Activities

Fewer games and activities are included in this chapter as students will be using measuring tools. The included activities can be used after students complete the **Do** questions, or anytime additional practice is needed.

Chapter Opener

Objective
- Investigate area.

Lesson Materials
- Rectangular towels or blankets of different but similar sizes

Bring in two blankets that are two different sizes; the differences in size should be so minimal that students cannot tell by just looking. Spread the blankets out in different parts of the classroom. Have students think about how they can compare the sizes of the blankets.

Discuss students' ideas. For example:

- Lay the blankets on top of one another.
- Measure the lengths and widths of each blanket.
- Find out how many people can sit on each blanket.
- Find out how many math textbooks can be placed on each blanket.

Laying the blankets on top of one another is a direct comparison. Having kids sit on them or tiling the blankets with math textbooks is the informal understanding of using units to find area.

Using student bodies is not very accurate. Students are different sizes and will be situated on blankets differently. For example, some might not have a full space or there may be a lot of space between students. Discussion from this activity will help prepare students to think about finding more efficient and accurate ways of finding area in the upcoming lessons.

Continue the discussion in Lesson 1 by having the students think about the given square and rectangle as possible options for the picnic blankets.

Lesson 1 Area

Objective
- Find the area of a figure in square units.

Lesson Materials
- Square and Rectangle (BLM)
- 1-in square tiles or pieces of paper
- Scissors

Think

Provide students with the Square and Rectangle (BLM) and 1-in square tiles. Pose the **Think** problem and have students try to solve the problem independently.

Discuss student methods for finding which paper shape covers more surface.

Learn

Have students work with a partner, so each pair has 2 sets of the Square and Rectangle (BLM).

Have one partner cut out the square and rectangle and line up the figures. They should then cut them similar to Dion's example. Students should note that the piece cut off from the square figure covers more space than the piece cut off from the rectangle.

If they continue to compare the two cut-off pieces, they can find the exact amount of area difference between the purple square and yellow rectangle, which is 1 square tile.

Have the other partner lay the square tiles over the square and rectangle as in Emma's example.

Have students count the square tiles to see which shape takes more square tiles to cover. Introduce the term "area" and discuss square units. Make sure students make the connection between the square tile and the square unit.

Students may relate the images to ones of arrays they have worked with in multiplication chapters from Dimensions Math 2A, 2B, and 3A.

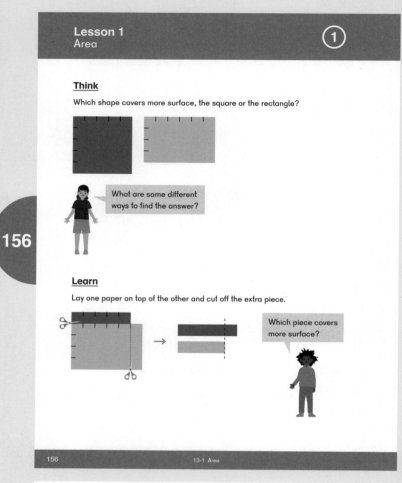

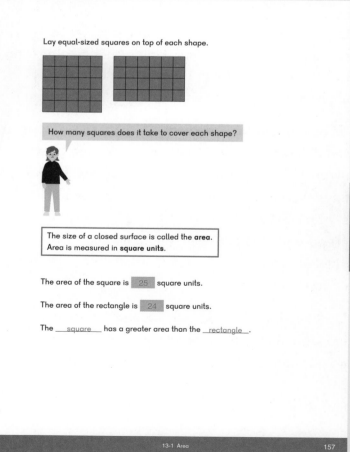

Do

❶ – ❷ Have students use the square tiles.

❸ Have students cut 2 square pieces of paper diagonally to make right triangles. Have them use all 8 pieces to make shapes and record the different shapes on grid paper.

This problem introduces a half-square. Just as two halves of a square can be put together to form the whole square, the two half-units can be put together to form 1 whole unit.

Students should note that even though the areas of the shapes are not nicely formed, they still have an area of 6 square units.

Some students may make shapes like this:

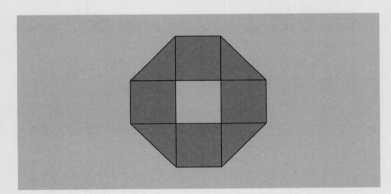

The area of the shape is still 6 square units.

Activity

▲ **Same Area**

Materials: 2 sets of square and half-square cards from ❸

Students work with the two sets of cards to make a rectangle and a triangle with the same area. Challenge them to use all 16 cards to make the two shapes with the same area.

Exercise 1 • page 143

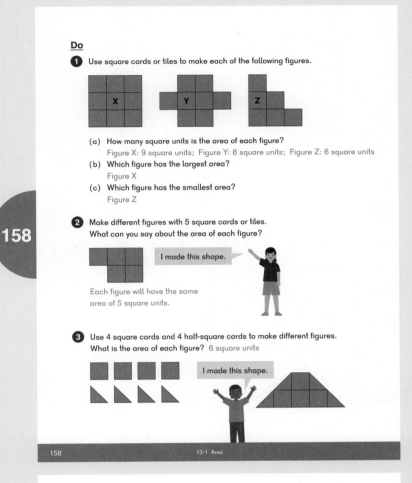

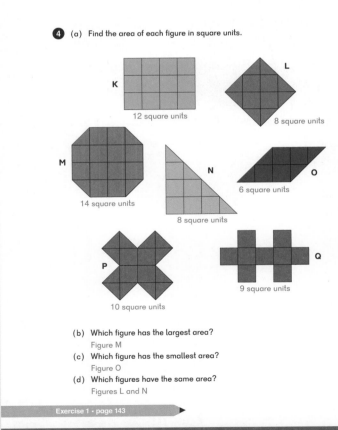

Lesson 2 Units of Area

Objectives

- Understand formal square units.
- Find the area of figures in formal square units.

Lesson Materials

- Centimeter Graph Paper (BLM)

Think

Provide students with Centimeter Graph Paper (BLM) and have them draw the shapes in **Think**.

Pose the **Think** questions and have students try to answer them independently.

Ask students how this problem is the same or different from the previous lesson. They should note that the square units in **Think** are square centimeters.

Learn

Introduce the notation for centimeter square units, cm², and have students write the area of the four shapes as cm². Students should read cm² as "square centimeters."

Discuss which shapes have the largest area, the smallest area, and the same areas.

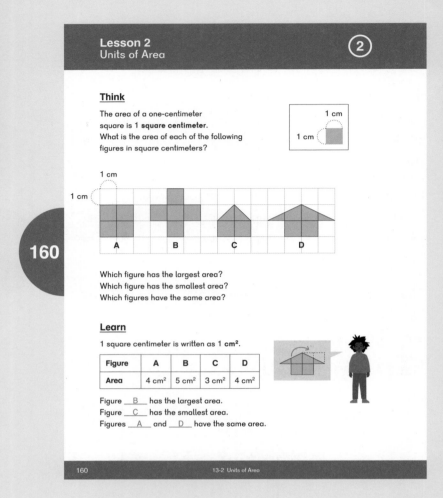

Do

① Have students share their drawn shapes. Ensure that they are counting the units correctly.

② If students come up with an answer that is not a whole number of square units, they have counted incorrectly.

③ — ⑤ These problems show other units of measurement as square units:

- 1 square inch is written 1 in^2
- 1 square foot is written 1 ft^2
- 1 square yard is written 1 yd^2
- 1 square meter is written 1 m^2
- 1 square mile is written as 1 mi^2

Discuss Sofia and Alex's questions.

We typically measure the area of a rug or the floor of a room in square feet or meters. We can measure the amount of paint that covers a wall in square feet.

We typically measure fabric in square yards.

We can measure the area of a state or city in square miles (mi^2).

Exercise 2 • page 146

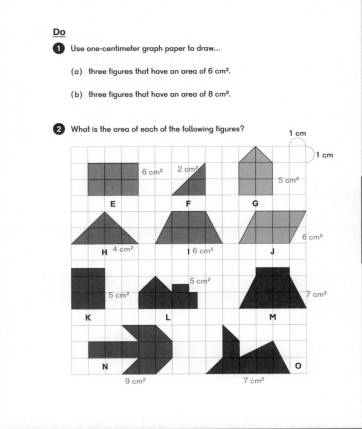

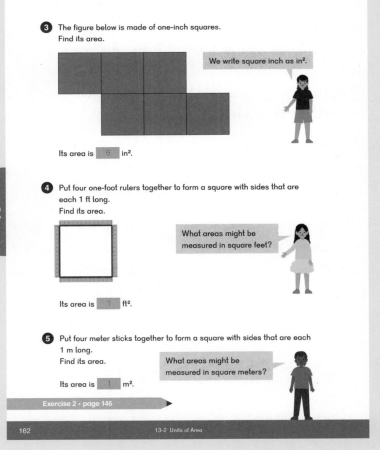

Lesson 3 Area of Rectangles

Objective

- Use multiplication to find the areas of rectangles and squares.

Lesson Materials

- 4" × 6" index cards
- One-inch square tiles, 12 per student

Think

Provide each student with an index card and twelve 1-inch square tiles, and pose the **Think** problem. Note that students will not have enough tiles to completely cover the index card. They should find an alternate strategy.

Have students try to solve the problem independently, then discuss their methods for finding the area of the card in square inches.

Learn

Discuss Dion and Emma's questions. Students may use the tiles to draw the marks to show 1-inch increments along the sides of the index card. The goal is to see that instead of counting all of the squares, we can multiply the number of 1-inch increments on the length by the width to find the area.

Students may recall that they could find the number of items in an array by multiplying the number of rows by the number of columns or vice versa.

If the postcard was completely covered with squares, the squares would form an array. The length of each side of the array corresponds to the number of 1-inch increments along each side.

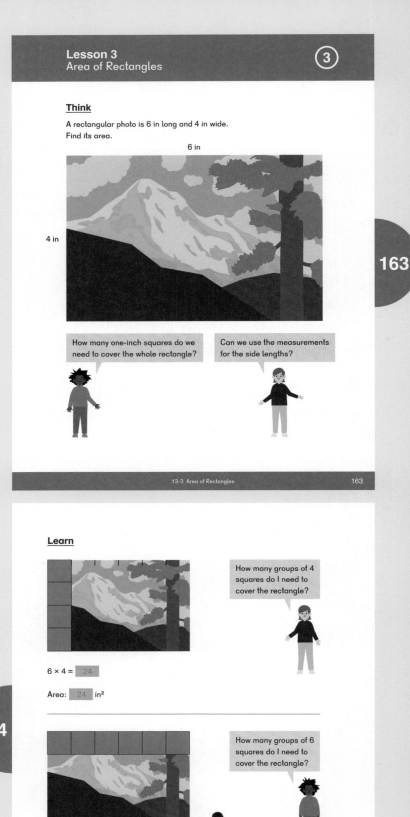

Do

① Students will need 12 one-inch square tiles.

② Students can work in small groups to make squares using more than 4 square tiles.

For example:
- 3 × 3 or 9 square units
- 4 × 4 or 16 square units
- 5 × 5 or 25 square units

Activity

▲ Fences

Materials: Centimeter Graph Paper (BLM) in a dry erase sleeve, pair of 10-sided dice, 2 colors of dry erase markers

Students take turns rolling the dice. On each turn, they fence in land on the Centimeter Graph Paper (BLM) by outlining an array as determined by the dice. For example, if a player rolls 3 and 4, she would make a 3 × 4 array.

The players write two multiplication equations on their newly acquired land. In the example shown here:

- Player One (red) rolled 6 and 7, and fenced in a 6 × 7 array.
- Player Two (blue) rolled 9 and 4, and fenced in a 9 × 4 array.

Play ends when a player cannot place an appropriate sized rectangle on the board. Each player adds up the total area they have fenced in. The player with the most area is the winner.

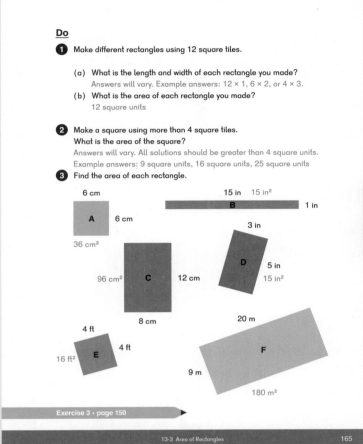

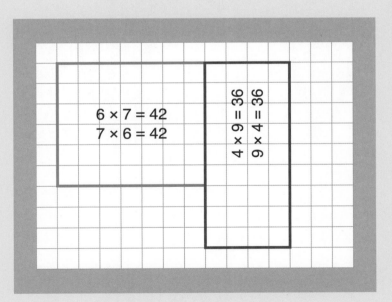

Exercise 3 • page 150

Lesson 4 Area of Composite Figures

Objective

- Find the area of composite shapes formed of rectangles.

Lesson Materials

- Area of Composite Figures (BLM)
- Centimeter Graph Paper (BLM)

Think

Provide each student with an Area of Composite Figures (BLM) and a Centimeter Graph Paper (BLM). Have them cut out and use the shapes to find a solution to Alex's question. They might fold or cut the shapes to find the area.

Discuss student methods for finding the area. Students should note that unlike previous lessons, they cannot simply multiply the length of the shape by its width.

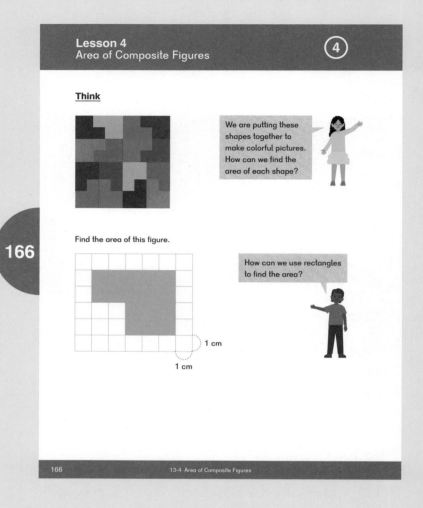

Learn

Dion and Sofia partition their shapes into two smaller rectangles. They find the area of the smaller rectangles and add them together to find the area of the shape.

Mei uses a different strategy. She imagines a whole rectangle and subtracts the missing part.

Students could also partition the shape into three smaller rectangles:

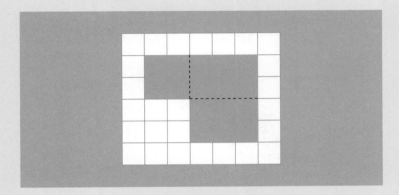

Regardless of the method used, students should see that to find the area of a composite shape, they can partition the figure into smaller rectangles. They can then multiply the length by the width to find the area of each smaller rectangle.

Students can add the areas together to find the total area.

Ensure that students understand that by cutting, or partitioning the shape, the area of the total shape does not change.

Have students compare their solutions from **Think** with the ones shown in the textbook.

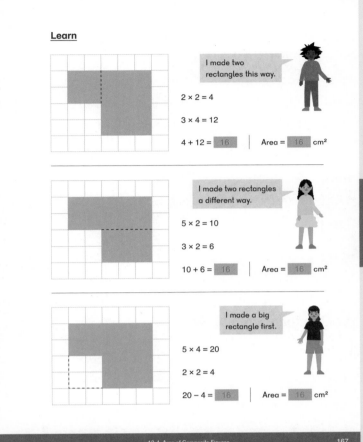

Do

1. Students may partition these shapes in different ways. While students can count the underlying squares to find the total area, encourage them to partition the shapes and find the areas of the smaller rectangles by multiplying the sides.

   Note that the answer overlay includes only one method for each figure. Students may use different methods and may use a combination of methods, as shown for D.

2. Some sides are not labeled. Ask students how they can find those lengths. For example, they may answer, "I can subtract the length of the shorter side from the length of the longer side."

2. (b) The figure provides enough information to also partition the shape in other ways:

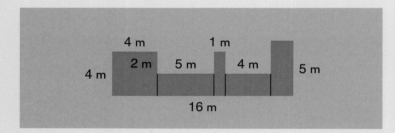

4. Provide students with Centimeter Graph Paper (BLM) and have them draw or cut out shapes to see how both Emma and Alex's methods work. Emma shifts a part of the shape. Alex doubles the area by adding a second shape of the same size. He then divides by 2 to find the area of one shape.

Exercise 4 • page 153

Do

1. Find the area of each figure.   Methods may vary.

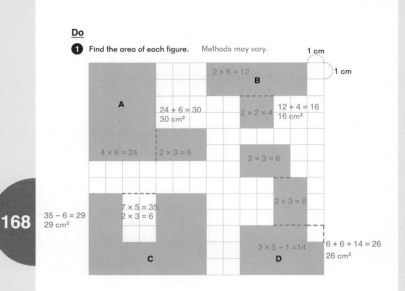

2. Each figure is made up of rectangles. Find the area of each figure.

   (a)
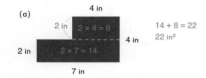

Methods may vary.

   (b)
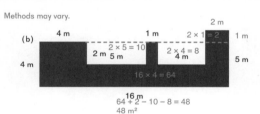

3. Mei has 2 pieces of colored paper. Each has one side that is 9 in long. One piece has a side that is 12 in long and the other has a side that is 5 in long. She puts them together to make a rectangular poster. Show two different ways to find the total area.

   Method 1: 9 in × 12 in = 108 in²
   9 in × 5 in = 45 in²
   108 in² + 45 in² = 153 in²

   Method 2: 12 in + 5 in = 17 in
   17 in × 9 in = 153 in²

4. Why do Emma's and Alex's methods for finding the area of this figure work?

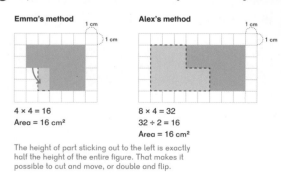

Emma's method
4 × 4 = 16
Area = 16 cm²

Alex's method
8 × 4 = 32
32 ÷ 2 = 16
Area = 16 cm²

The height of part sticking out to the left is exactly half the height of the entire figure. That makes it possible to cut and move, or double and flip.

Exercise 4 • page 153

Lesson 5 Practice A

Objective

- Practice finding the area of shapes.

After students complete the **Practice** in the textbook, have them continue to practice using activities from the chapter.

Activity

▲ **How Many?**

Materials: Square tiles

Have students make all possible composite figures using 3, 4, and 5 squares. Squares next to each other need to line up along the entire lengths or sides.

How many different shapes can they make? Can they find them all? (There are two shapes for 3-unit squares, five shapes for 4-unit squares, and 12 for 5-unit squares.)

Exercise 5 • page 157

Lesson 6 Perimeter

Objectives

- Understand the meaning of perimeter.
- Find the perimeter of polygons.

Lesson Materials

- 1-inch square tiles
- Centimeter ruler

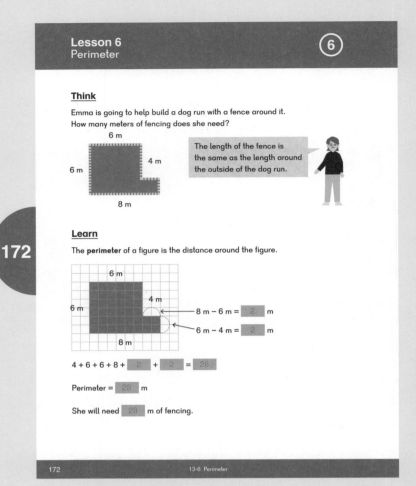

Think

Pose the **Think** problem and discuss Emma's comment.

Have students try to solve the problem independently. They should notice that they are missing some side lengths. Ask them how they can find those lengths.

Learn

Introduce the term "perimeter." A shape's perimeter is distance around the shape.

Students should see that to find the perimeter of a shape, they need to find the total length around the outside of the shape.

Point out to students that since perimeter is a measure of length, we measure in units of length, not square units. Perimeter can be measured in many units: centimeters, meters, inches, feet, and other units of length.

Have students compare their solutions from **Think** with the one shown in the textbook.

Do

1. Ensure students are finding the length around the figure (perimeter), not the number of tiles that make up the figure (area).

2. For Figure F, have students estimate, then measure the length of the long side of the triangle with a centimeter ruler. (10 cm)

3. If students struggle, remind them that they need to find the lengths of sides that are not given, not merely add together the numbers provided.

4. As the corners do not overlap, students find the width by subtracting 1 in from each side: 5 − 1 − 1 = 3.

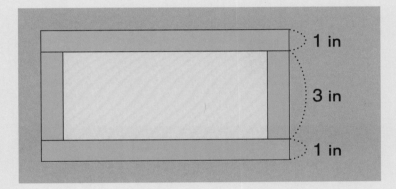

They can then add 9 + 3 + 9 + 3 to find the perimeter of 24 in.

Exercise 6 • page 161

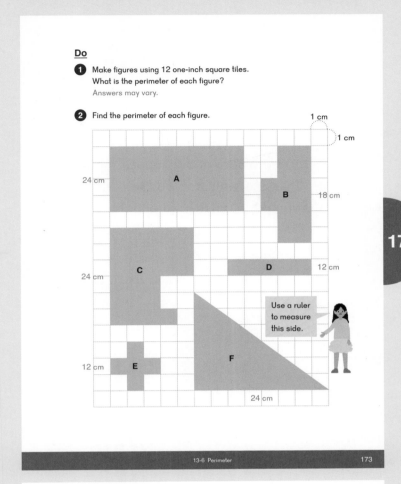

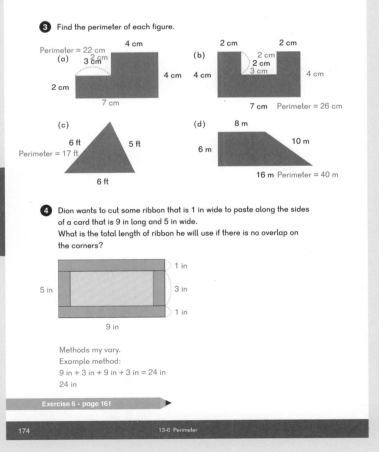

Lesson 7 Perimeter of Rectangles

Objective

- Calculate the perimeter of rectangles given the length and width.

Think

Pose the **Think** problem and discuss Emma's comment. While students can add to check their work, they should try to use multiplication to find the perimeter of the shapes. Have students try to solve the problem independently.

Discuss student methods for finding the perimeter of the shapes.

Learn

Help students see that in the rectangle there are two sets of 9 in + 5 in, so we can add the length and width and then multiply the result by 2.

Students should also see that two sides are 9 inches and two sides are 5 inches. We can multiply 9 in × 2 and 5 in × 2 and add the products.

The perimeter of the square can be found by multiplying the side length × 4.

Have students compare their solutions from **Think** with the one shown in the textbook.

In the **Do** section, students will find the perimeter and the area of a rectangle. Prepare for this by finding the area of the rectangles in **Think**.

Do

Ensure students understand the difference between perimeter and area. Remind them that perimeter is the total length around the outside of a figure, and the area is the amount of surface area inside the figure.

Exercise 7 • page 164

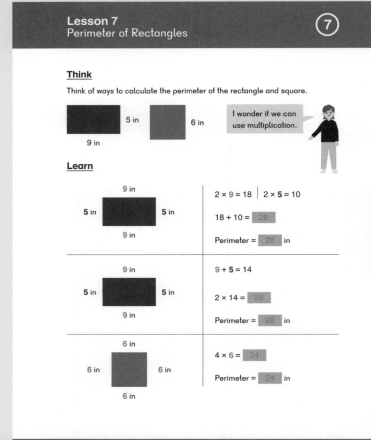

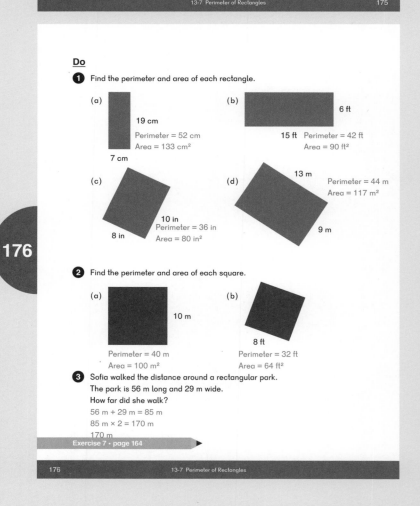

Lesson 8 Area and Perimeter

Objective
- Understand the relationship between area and perimeter.

Lesson Materials
- Toothpicks or craft sticks, 12 per student
- Centimeter Graph Paper (BLM)

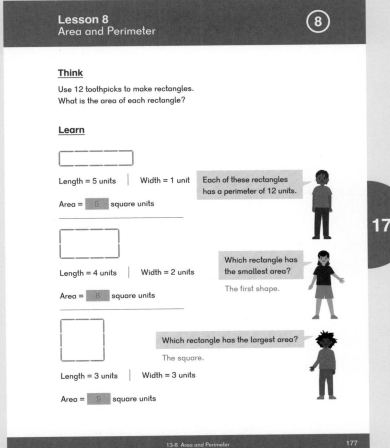

Think

Provide each student with 12 toothpicks and have them make different rectangles using all of the toothpicks. The toothpicks should not overlap or be built inside of another rectangle.

Students can record their shapes and calculations on Centimeter Graph Paper (BLM).

Learn

Alex, Mei, and Dion all show different rectangles with a perimeter of 12 units.

Mei asks which rectangle has the smallest area. Students should see that the rectangle with a width of 1 unit has the smallest area. Given a fixed perimeter, the shape with the largest area is the shape that is closest to a square. This might seem counterintuitive to students.

Have students compare their rectangles and areas with the ones shown in the textbook.

© 2017 Singapore Math Inc.   Teacher's Guide 3B Chapter 13

Do

1. Students will now use graph paper and rectangles with an area of 12 cm² to find different perimeters.

2—4 Students will find both the area and perimeter to answer the questions.

4. Each figure in 4 (a) removes a square from a corner of the previous shape. While the area changes, the perimeter does not.

In 4 (b), both the area and perimeter change when squares are removed from the middle of the shape.

Activity

★ **Toothpick Puzzler**

Materials: 12 toothpicks

Create a shape where the number of toothpicks, or perimeter, is exactly double the area.

Answer:

Create a shape where the number of toothpicks is more than double the area.

Answers will vary.

Exercise 8 • page 167

Do

The rectangles will have the following dimensions: length = 6 cm, width = 2 cm; length = 2 cm, width = 6 cm; length = 3 cm, width = 4 cm; length = 4 cm, width = 3 cm; length = 1 cm, width = 12 cm; and length = 12 cm, width = 1 cm.

1. Use centimeter graph paper to draw all the possible rectangles with side lengths equal to a whole number and an area equal to 12 cm².

   (a) What is the longest possible perimeter?
       26 cm
   (b) What is the shortest possible perimeter?
       14 cm

2. These figures are made up of one-centimeter squares.

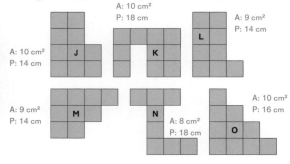

   (a) Which figures have the same area but different perimeters?
       Figures J, K, and O.
   (b) Which figures have the same perimeters but different areas?
       Figures K and N, figures J and L, figures J and M.
   (c) Which figures have the same area and perimeter?
       Figures L and M.

3. Mei and Alex are helping to put a fence around a garden. They want the garden that has the most space and the least fencing. Which of the following gardens should they choose and why?

They should choose Garden B because it has a greater area and a shorter perimeter than Garden A.

4. Compare the area and perimeter of these figures. What do you notice?

(a)

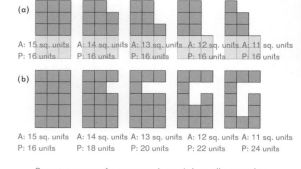

A: 15 sq. units   A: 14 sq. units   A: 13 sq. units   A: 12 sq. units   A: 11 sq. units
P: 16 units   P: 16 units   P: 16 units   P: 16 units   P: 16 units

(b)

A: 15 sq. units   A: 14 sq. units   A: 13 sq. units   A: 12 sq. units   A: 11 sq. units
P: 16 units   P: 18 units   P: 20 units   P: 22 units   P: 24 units

Removing a square from a corner does not change the perimeter.
Removing a square from a side adds 2 units to the perimeter.

Exercise 8 • page 167

Lesson 9 Practice B

Objective

- Practice finding area and perimeter.

After students complete the **Practice** in the textbook, have them continue to practice by playing games from this chapter.

Activity

▲ **Area Game**

Materials: Centimeter Graph Paper (BLM) in a dry erase sleeve, a pair of 6-sided dice, 2 colors of dry erase markers

This is an extension of **Fences** from Lesson 3.

Players take turns rolling the dice. On each turn, they roll the dice, multiply the numbers, and shade an equivalent area on the Centimeter Graph Paper (BLM).

In the example shown below:

- Player One (blue) rolled a 5 and 2, and fenced an area of 10 square units.
- Player Two (red) rolled 6 and 4, and fenced in an 8 × 3 array, or 24 square units.

Play ends when a player cannot place an appropriate sized shape on the board. Each player adds up the total amount of area they have fenced in. The player with the most area is the winner.

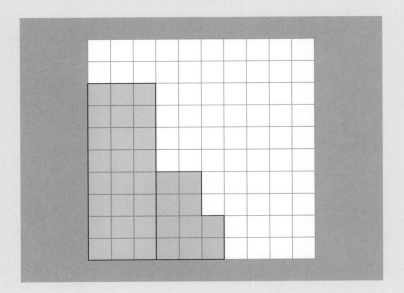

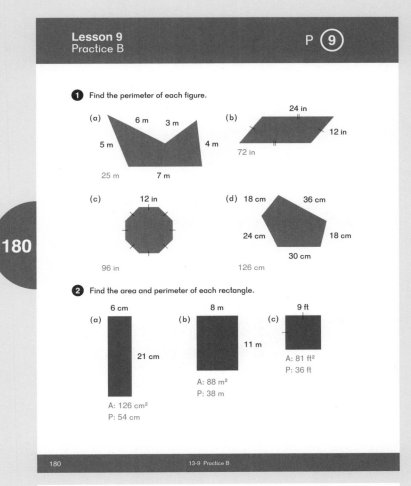

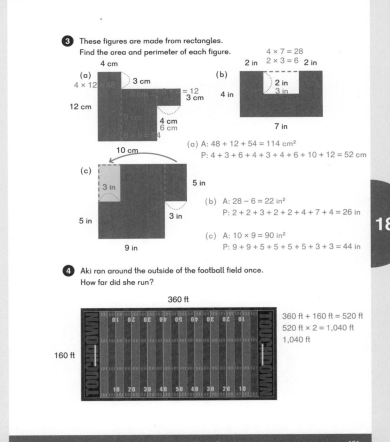

Exercise 9 • page 170

Brain Works

★ Tiling

Materials: Inch Graph Paper (BLM) or graph paper

For this independent activity, provide students with Inch Graph Paper (BLM) or graph paper and the following situation:

"On a 10-square unit grid, you can lay rectangles that are 2 square units, 3 square units, and 4 square units. How many different ways can you cover the grid?"

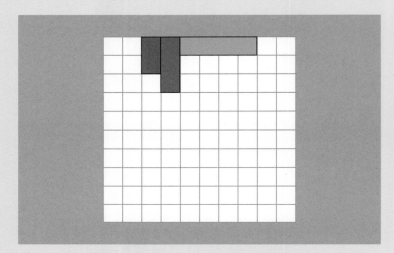

Challenge students by asking, "On a 10-square unit grid, you can lay shapes that are 2 square units, 3 square units, and 4 square units. How many different ways can you cover the grid?"

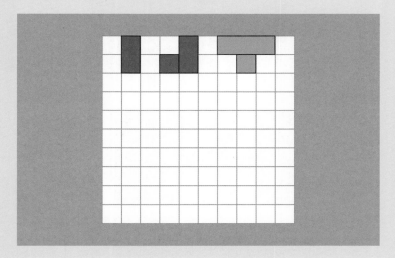

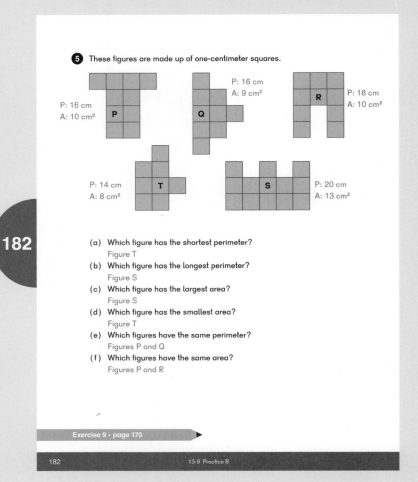

Exercise 1 • pages 143–145

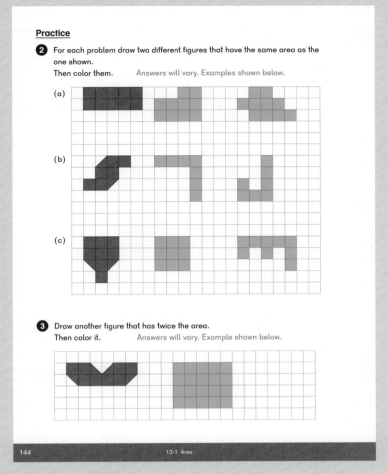

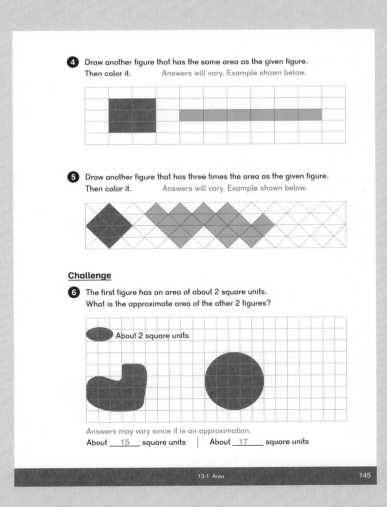

Teacher's Guide 3B Chapter 13

Exercise 2 • pages 146–149

Exercise 2

Basics

1 (a) Find the area of each figure in square centimeters.

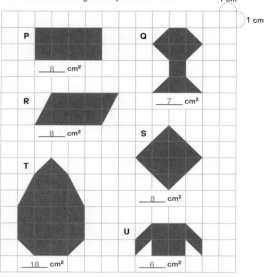

P ___8___ cm²
Q ___7___ cm²
R ___8___ cm²
S ___8___ cm²
T ___18___ cm²
U ___6___ cm²

(b) Figure ___T___ has the largest area.

(c) Figure ___U___ has the smallest area.

(d) Which figures have the same area?
P, R, S

(d) 6 cm² larger.

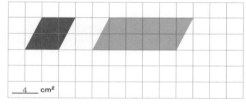

___4___ cm²

(e) 2 cm² smaller.

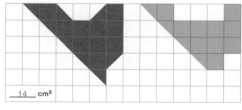

___14___ cm²

(f) 5 cm² smaller.

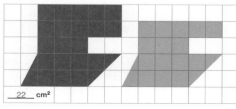

___22___ cm²

Practice

2 Each square on the grid below has an area of 1 cm².
Write the area of each figure, then draw another figure that is...
Drawings will vary. Examples shown below.

(a) the same area.

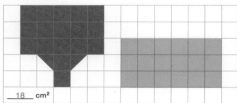

___18___ cm²

(b) the same area.

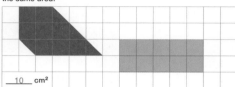

___10___ cm²

(c) 3 cm² larger.

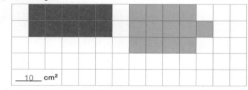

___10___ cm²

3 Each square is 1 inch long.
The area of the shaded figure is ___5 in²___.

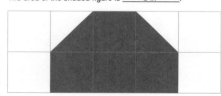

4 Fill in the blanks with in², m², or ft².

(a) The floor of a bedroom has an area of about 80 ___ft²___.

(b) A piece of paper has an area of about 94 ___in²___.

(c) A car parking space is about 12 ___m²___.

Challenge

5 Each square has an area of 1 cm².
What is the area of each triangle?

(a)  2 cm²

(b) 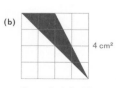 4 cm²

One method: Cut the shaded part below the middle horizontal line, rotate and move it. The figure now covers half of 4 squares.

One method: Find the area of the unshaded part and subtract from the total.

Exercise 3 • pages 150–152

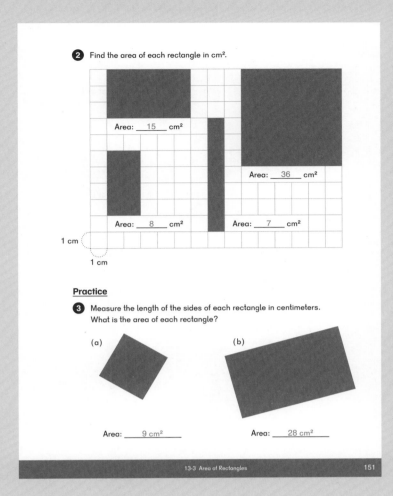

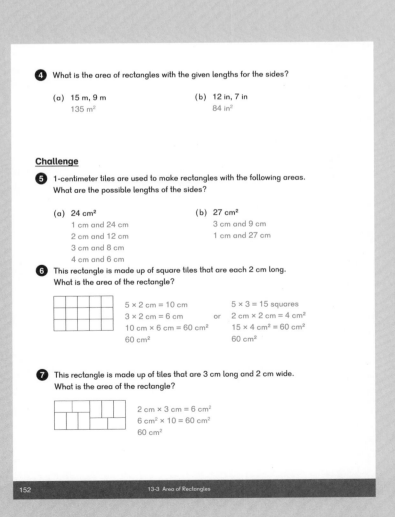

Exercise 4 • pages 153–156

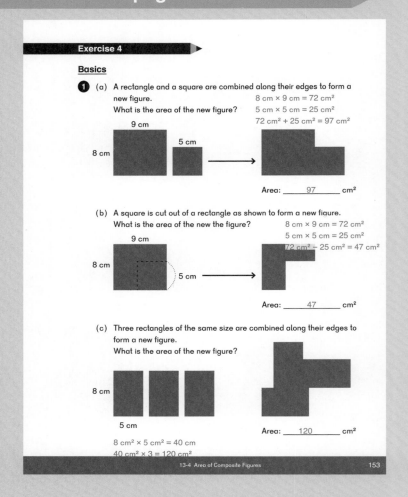

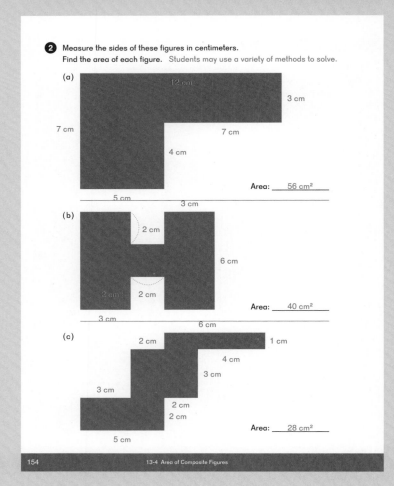

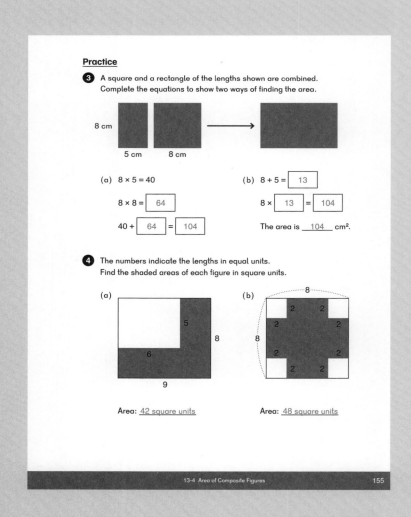

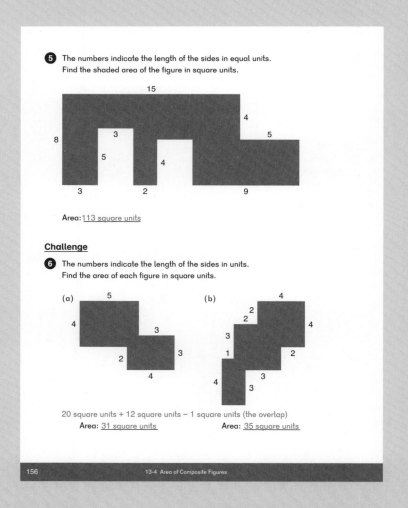

Exercise 5 • pages 157–160

Exercise 5

Check   Drawings will vary. Examples shown below.

1. Draw a figure with an area that is 7 square units more than the area of the given figure.

2. Draw two different figures with areas of...

   (a) 17 square units.

   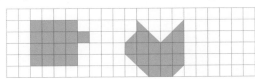

   (b) 23 square units.

   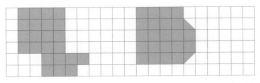

3. Three rectangles are 2 cm by 2 cm, 4 cm by 3 cm, and 5 cm by 3 cm. Draw a figure showing the 3 rectangles joined along the edges with no overlap. What is the area of the figure?
   Drawings will vary. Example shown below.

   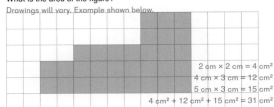

   2 cm × 2 cm = 4 cm²
   4 cm × 3 cm = 12 cm²
   5 cm × 3 cm = 15 cm²
   4 cm² + 12 cm² + 15 cm² = 31 cm²

4. Draw two different rectangles with areas of 16 cm².
   Drawings will vary. Examples shown below.

   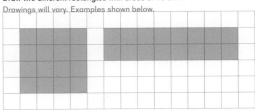

5. A rectangular sign is 32 cm long and 8 cm wide. What is the area of the sign?
   32 cm × 8 cm = 256 cm²
   The area of the sign is 256 cm².

6. A rectangular piece of plywood is 4 ft wide. It is 3 times as long as it is wide. What is the area of the piece of plywood?
   4 ft × 3 = 12 ft (length)
   12 ft × 4 ft = 48 ft²
   The area of the plywood is 48 ft².

7. A piece of cardboard is 54 cm long. It is 6 times as long as it is wide. What is the area of the piece of cardboard?
   54 cm ÷ 6 = 9 cm (width)
   54 cm × 9 cm = 486 cm²
   The area of the piece of cardboard is 486 cm².

8. A square piece of paper has an area of 36 cm². A smaller square measuring 3 cm on one side is cut from the paper. What is the area of the remaining piece of paper?
   3 cm × 3 cm = 9 cm²
   36 cm² − 9 cm² = 27 cm²
   The area of the remaining piece of paper is 27 cm².

9. The cost of installing a certain carpet is $8 per square foot. How much would it cost to carpet a room with the following dimensions?

   Students may use another strategy.
   27 ft × 9 ft = 243 ft²
   3 ft × 12 ft = 36 ft²
   243 ft² − 36 ft² = 207 ft² (total area to carpet)

   $8 × 207 = $1,656
   It would cost $1,656 to carpet the room.

Challenge

10. Two squares with side lengths of 10 cm overlap by 4 cm. What is the area of the resulting rectangle?

    Area of each square is 100 cm².
    Area of the overlap is 40 cm².
    100 cm² + 100 cm² − 40 cm² = 160 cm²
    or
    New double length: 10 cm + 10 cm − 4 cm = 16 cm
    16 cm × 10 cm = 160 cm²
    160 cm²

11. By how much should the following rectangles overlap in order to create a rectangle with an area of 144 cm²?

    108 cm² + 90 cm² = 198 cm²
    198 cm² − 144 cm² = 54 cm²
    54 cm² ÷ 9 cm = 6 cm
    or
    144 cm² ÷ 9 cm = 16 cm
    10 cm + 12 cm = 22 cm
    22 cm − 16 cm = 6 cm

    They should overlap by 6 cm.

12. Plastic rectangular tiles 2 cm wide and 3 cm long are being used to cover a piece of cardboard that is 10 cm by 13 cm. What is the greatest number of tiles that can be used?

    21 tiles
    Students can use logical trial and error on graph paper. Other arrangements of the tiles are possible.

Exercise 6 • pages 161–163

Exercise 6

Basics

1. A triangle has sides that are 5 cm, 6 cm, and 4 cm long. Find the perimeter of the triangle.

   $5 + 6 + 4 =$ 15

   Perimeter: __15__ cm

2. Measure the sides of these figures in centimeters and find their perimeters.

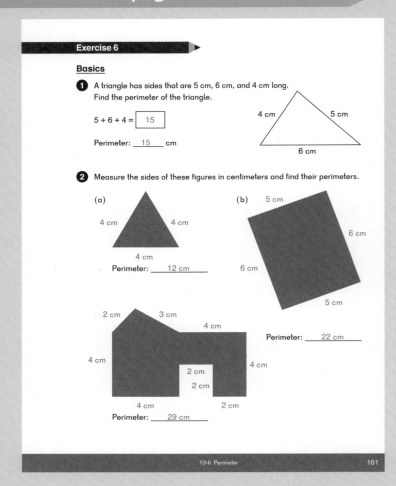

Practice

3. The lengths of the sides of each figure are marked in equal units. Find the perimeter of each figure in units.

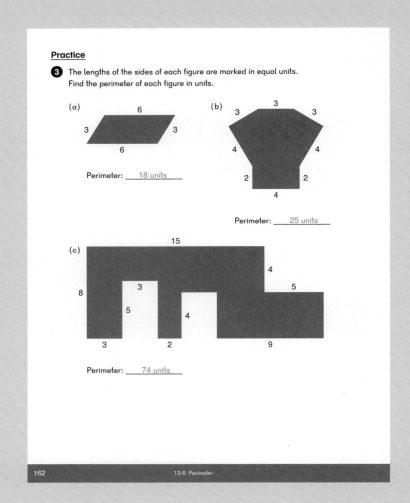

(a) Perimeter: __18 units__

(b) Perimeter: __25 units__

(c) Perimeter: __74 units__

4. A hexagon has 6 sides, each with a length of 12 cm. What is the perimeter of the hexagon?

   6 × 12 cm
   72 cm

5. The sum of the lengths of 3 sides of a rhombus is 84 cm. What is the perimeter of the rhombus?

   84 cm ÷ 3 = 28 cm
   28 cm × 4 = 112 cm
   112 cm

Challenge

6. The numbers indicate the length of the sides in units. Find the perimeter of each figure.

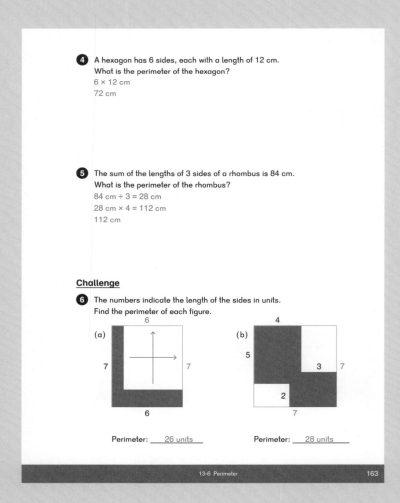

(a) Perimeter: __26 units__

(b) Perimeter: __28 units__

Exercise 7 • pages 164–166

Exercise 7

Basics

1. A rectangle is 9 cm long and 5 cm wide.
   Complete the equations to show two different methods to calculate the perimeter.

   9 cm
   5 cm

   (a) 9 + 5 = 14

   14 × 2 = ☐ 28

   (b) 9 × 2 = 18

   5 × 2 = ☐ 10

   18 + ☐ 10 ☐ = ☐ 28

   The perimeter is __28__ cm.

2. One side of a square is 7 cm long.
   Complete the equation to find the perimeter.

   7 cm

   4 × ☐ 7 ☐ = ☐ 28

   The perimeter is __28__ cm.

164 13-7 Perimeter of Rectangles

Practice

3. Find the perimeters of the following rectangles in units.
   The lengths of the sides are marked in equal units.

   (a)
   3
   5

   (b)
   6
   1

   Perimeter: __16 units__   Perimeter: __14 units__

4. Find the perimeters of the following squares in units.
   The lengths of a side is marked in equal units.

   (a)
   6

   (b)
   4

   Perimeter: __24 units__   Perimeter: __16 units__

5. A rectangular sign is 15 m long and 8 m wide.
   What is the perimeter of the sign?

   15 m + 8 m = 23 m
   23 m × 2 = 46 m
   The perimeter is 46 m.

13-7 Perimeter of Rectangles 165

6. A square piece of paper has a side of 18 cm.
   What is its perimeter?
   18 cm × 4 = 72 cm
   Its perimeter is 72 cm.

7. A square has a perimeter of 40 cm.
   What is the length of one side?
   40 cm ÷ 4 = 10 cm
   One side is 10 cm long.

Challenge

8. A rectangle has a perimeter of 30 cm.
   One side is 8 cm long.
   What is the length of the other side?
   30 cm ÷ 2 = 15 cm
   15 cm − 8 cm = 7 cm
   The other side is 7 cm long.

9. A square has a perimeter of 32 cm.
   It is cut into four equal squares as shown.
   What is the perimeter of one of the small squares?
   32 cm ÷ 4 = 8 cm
   8 cm ÷ 2 = 4 cm
   4 cm × 4 = 16 cm
   The perimeter of a small square is 16 cm.

166 13-7 Perimeter of Rectangles

© 2017 Singapore Math Inc.   Teacher's Guide 3B Chapter 13

Exercise 8 • pages 167–169

Exercise 8

Basics

1 Each small square on the grid has a side length of 1 unit.

K — A: 26 square units, P: 30 units
L — A: 23 square units, P: 24 units
M — A: 28 square units, P: 32 units
N — A: 30 square units, P: 26 units
O — A: 30 square units, P: 26 units
P — A: 26 square units, P: 24 units

(a) Which figures have the same area and different perimeters?
K and P
(b) Which figures have the same perimeter but different areas?
L and P
(c) Which figures have the same area and perimeter?
N and O

Practice

2 Use the grid to draw three different rectangles with perimeters of 16 units. Write the area of each figure.
Which figure has the smallest area?
Drawings will vary. Examples shown below.

Area = 16 square units
Area = 12 square units
Area = 7 square units — smallest area

3 Shade one square so that the area of the figure is increased by 1 square unit, but the perimeter...

(a) decreases.
(b) stays the same.
(c) increases.

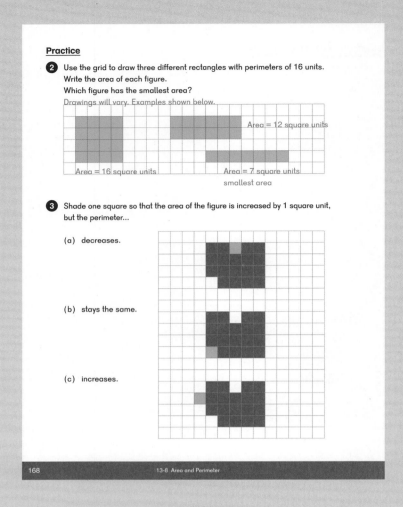

4 Draw another figure that has... Drawings will vary. Examples shown below.

(a) the same area and perimeter.

(b) a greater perimeter but smaller area.

Challenge

5 Draw a rectangle with the largest possible perimeter that has a smaller area than the given figure.
The sides of the rectangle must be whole numbers of units.

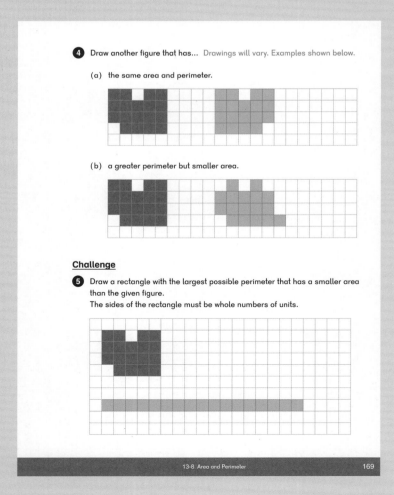

Exercise 9 • pages 170–172

Exercise 9

Check

1. The following figures are made up of square units.
   The numbers indicate the length of the sides in units.
   Find the area and perimeter of each figure.

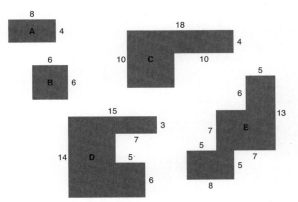

	Area	Perimeter
A	32 square units	24 units
B	36 square units	24 units
C	120 square units	56 units
D	163 square units	68 units
E	140 square units	66 units

2. A rectangle has a length of 16 cm and a width of 9 cm.
   What is its area and perimeter?
   16 cm × 9 cm = 144 cm²
   Area = 144 cm²
   16 cm + 9 cm = 25 cm
   25 cm × 2 = 50 cm
   Perimeter = 50 cm

3. The diameter of each circle is 6 units.
   What is the perimeter of the quadrilateral?

   Radius = 3 units
   5 × 3 = 15
   15 units

4. Draw a figure with an area of 35 square units and a perimeter of 26 units.
   Drawings will vary. Example shown below.

Challenge

5. A square is cut into four equal squares as shown.
   The perimeter of each smaller square is 12 cm.
   What is the area of the larger square?

   12 cm ÷ 4 = 3 cm
   3 cm + 3 cm = 6 cm (length of one side of larger square)
   6 cm × 6 cm = 36 cm²
   36 cm²

6. The figure below is made up of three identical squares, each with a side of 6 cm.
   What is the perimeter of the figure?

   12 cm × 4 = 48 cm
   48 cm

7. A rectangular piece of paper is folded to form the shape shown below.
   What is the perimeter and area of the piece of paper before it is folded?

   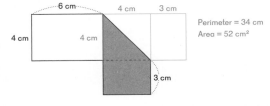

   Perimeter = 34 cm
   Area = 52 cm²

Notes

Chapter 14 Time — Overview

Suggested number of class periods: 7–8

	Lesson	Page	Resources	Objectives
	Chapter Opener	p. 213	TB: p. 183	Investigate elapsed time.
1	Units of Time	p. 214	TB: p. 184 WB: p. 173	Review the analog clock, telling time, and the meaning of a.m. and p.m. Convert hours and minutes to minutes only and vice versa. Convert minutes and seconds to seconds only and vice versa.
2	Calculating Time — Part 1	p. 218	TB: p. 190 WB: p. 177	Find elapsed time within a.m. or p.m. Find an end time given a start time and elapsed time, within a.m. or p.m. Add time in hours and minutes, using mental math strategies for "making 60."
3	Practice A	p. 220	TB: p. 194 WB: p. 180	Practice concepts from the first two lessons.
4	Calculating Time — Part 2	p. 221	TB: p. 196 WB: p. 183	Add periods of time in hours and minutes where the time changes from a.m. to p.m. or from p.m. to a.m. Find end times, when given start and elapsed times.
5	Calculating Time — Part 3	p. 224	TB: p. 200 WB: p. 187	Subtract periods of time in hours and minutes that do not cross a.m. or p.m. Find start times, when given the end times and elapsed time.
6	Calculating Time — Part 4	p. 227	TB: p. 204 WB: p. 191	Subtract periods of time in hours and minutes that cross a.m. and p.m.
7	Practice B	p. 229	TB: p. 207 WB: p. 195	Practice concepts from the chapter.
	Workbook Solutions	p. 230		

Chapter 14 Time

Notes

In Dimensions Math 2B Chapter 12: Time, students learned to:

- Tell time to the minute.
- Determine intervals of time in minutes within one hour.
- Determine intervals of time in hours.
- Find the end time, when given the start time and elapsed time.
- Find the start time, when given the end time and elapsed time.

In this chapter, students will learn to convert units of time between hours, minutes, and seconds, as well as between days and weeks. Understanding units of time can be challenging for students. Since grade 1, they have been regrouping quantities in base 10 only. Now, they will have to regroup by 60.

For Chapter Opener and Lesson 1, clocks with geared hands where the hour hand moves in conjunction with the minute hand, not separately (often called "Judy Clocks"), are recommended.

In Lessons 2 through 7, students will calculate elapsed time by extending strategies from mental math methods introduced in Dimensions Math 3A Chapter 2: Addition and Subtraction — Part 1. They will also find start or end times when given start (when asked for end) or end (when asked for start) times and elapsed time.

Dimensions Math books include this emphasis on time and elapsed time as an application of mental math skills. Calculating elapsed time or determining start and end times requires mentally regrouping by 60, a skill that reinforces more abstract regrouping skills beyond tens.

Timelines

The use of a timeline when calculating elapsed time can be helpful. Have students draw tick marks and units of time to help them solve the problems.

Example: Dion went to bed at 9:45 p.m. He slept for 8 h 15 min. What time did he wake up?

Method 1

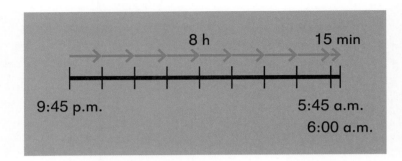

The student using Method 1 draws a timeline and writes 9:45 p.m. as a start time. She then counts on hours to 5:45 a.m., then in minutes to 6 a.m. The time intervals and spaces will be approximate. If needed, students can use a Timeline (BLM).

Method 2

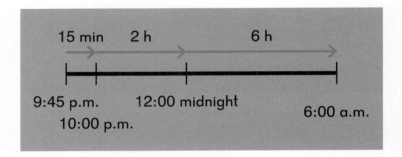

Students could also break up times into different intervals. The student using Method 2, as shown above, counts on 15 minutes first to get to 10:00 p.m., then 2 hours to 12:00 a.m. (midnight), and finally 6 more hours to 6:00 a.m.

Method 3

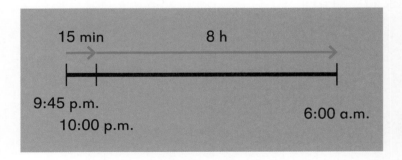

The student using Method 3, as shown above, also counts on 15 minutes first to 10:00 p.m., then 8 hours to 6:00 a.m.

Chapter 14 Time

Teachers should incorporate time into daily activities.

For example:

- It is time to start math class. What time is it?
- We are going to gym class. What time is it?
- Math class ends in one hour. What time will it be when math class ends?
- How much time before we go to gym class?

A.M. and P.M.

Most digital clocks designate midnight as 12 a.m. and noon as 12 p.m. The terms "midnight" and "noon" are generally considered less confusing designations and will be used in this series.

Students may bring up the idea of a 24-hour clock, also referred to as international time or military time. In 24-hour notation, the day begins at midnight or 00:00 and time is read in the hundreds. A 24-hour clock is commonly used in the sciences, medical fields, and airlines to ensure there is no confusion between a.m. and p.m. times. The colon or period is used in some countries and fields, and omitted in others.

8:00 a.m. can be written 08:00, 0800, or 08.00 and is read, "Oh eight hundred."

8:00 p.m. can be written 20:00, 2000, or 20.00 and is read, "Twenty hundred."

Chapter 14 Time   Materials

Materials

- Clock with a second hand
- Demonstration clock
- Stopwatch or timer
- Student clocks with geared hands
- Whiteboards

Blackline Masters

- Match Time Cards
- Timeline

Activities

There are fewer games included in this chapter as students should be working with telling time and finding elapsed time. Use activities from previous chapters to continue to practice mental math strategies for addition and subtraction as well as multiplication and division facts for 2 through 10.

Chapter Opener

Objective
- Investigate elapsed time.

Lesson Materials
- Student clocks with geared hands

Review the hour hand and minute hand on a clock face. Discuss the smaller tick marks on a clock that denote 60 minutes in one hour.

Provide students with clocks and have them show the times that the five friends are sharing.

Encourage students to find the times of daily events that are important to them.

Ask questions such as:

- How much time do we have left in the lesson?
- How long is your ride to school?

Provide students with clocks and have them show the times within one hour, and within 6 hours. Ask questions such as:

- Set your clock to 4:15. Show the time that is 1 hour earlier than 4:15.
- Show a time that is about 12:30.
- Show a time that is a bit after 10:00.
- Set your clock to 11:50. Show the time that is 15 minutes later than 11:50.

Activity

▲ **Clock Nim**

Materials: Student clocks with geared hands

Beginning with the hands on the clock at 12:00, players take turns moving the minute hand clockwise 5, 10, or 15 minutes.

The player who moves the hands to 3:00 exactly is the winner.

Lesson 1 Units of Time

Objectives

- Review the analog clock, telling time, and the meaning of a.m. and p.m.
- Convert hours and minutes to minutes only and vice versa.
- Convert minutes and seconds to seconds only and vice versa.

Lesson Materials

- Student clocks with geared hands
- Stopwatch or timer

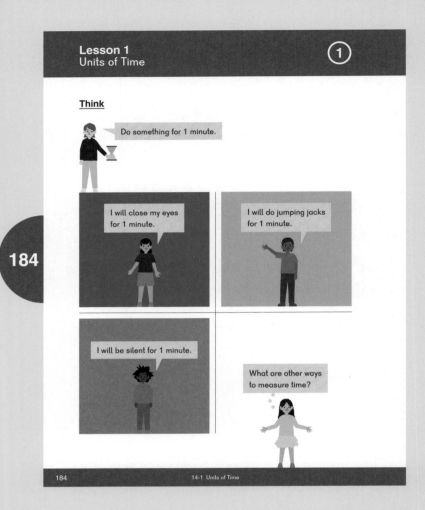

Think

Without looking at a clock, watch, or timer, have students try to estimate how long a minute is by keeping their heads down and eyes closed, then standing up when they think a minute has passed.

Repeat the activity using a timer.

Have students do an activity for 1 minute similar to the friends in **Think**. Discuss other measurements of time that students know:

- Seconds
- Hours
- Days
- Weeks
- Months
- Years

Learn

Discuss **Learn** and introduce the term "seconds." Ensure students understand that seconds are very short units of time. Help students see that just as there are 60 minutes in an hour, there are also 60 seconds in one minute.

Discuss Sofia's question regarding the third hand on the clock. Students should already know the minute hand and the hour hand.

Review a.m. and p.m. Discuss activities students do in the a.m. and p.m.

Most digital clocks designate midnight as 12:00 a.m. and noon as 12:00 p.m. The terms "midnight" and "noon" are generally considered less confusing terms and will be used in this series.

Do

1. Alex introduces the abbreviation of "s" for seconds. Students should recall the abbreviations "min" and "h" from Dimensions Math 2B. Note that students may see other abbreviations for seconds and hours, for example, "sec" or "hr."

2. (a)–(d) Students who are struggling can use a student clock.

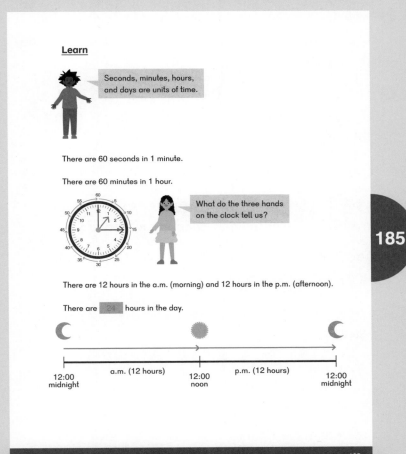

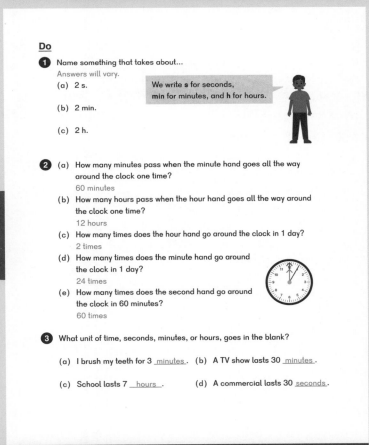

④—⑤ Dion realizes an hour is 60 minutes, and then can add parts (60 minutes and 30 minutes) to find the whole.

⑤ (a) Students can use Dion's strategy and think of 2 hours as 120 minutes by finding either 2 × 60 or 60 + 60.

(b) Add 5 minutes to the previous calculated answer.

⑥—⑦ Sofia uses the same strategy as Dion when converting minutes and seconds to seconds.

⑦ (a) Students can think of 2 minutes as 120 seconds by finding either 2 × 60 or 60 + 60.

(b) Add 55 seconds to the previously calculated answer.

(c) 1 min = 60 s
2 min = 2 × 60 s
3 min = 3 × 60 s
4 min = 4 × 60 s

⑧—⑩ Have students discuss Emma's thought as she converts minutes to hours and minutes. They can apply the same strategy to ⑩ when converting seconds to minutes and seconds.

⑪ Dion reminds students that there are 24 hours in one day. They can add 24 + 24 or multiply 24 × 2 before adding 4 more hours.

⑫ 1 day = 24 hours
2 days = 2 × 24 hours
3 days = 3 × 24 hours
...
7 days = 7 × 24 hours

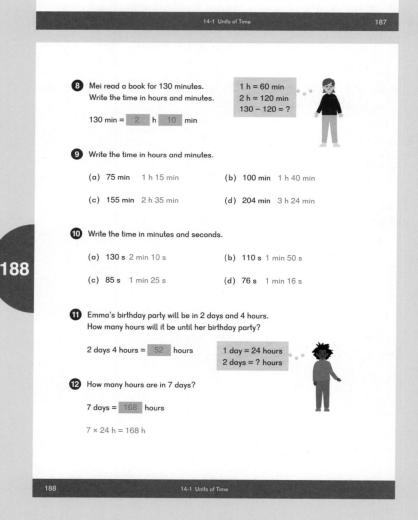

14 If needed, extend Mei's thought to show the progression:

1 week = 7 days
2 weeks = 2 × 7 days
3 weeks = 3 × 7 days

Continue to discuss other units of time:

1 year = 365 days
2 years = 2 × 365 days

Sofia prompts a discussion of the different number of days in each month.

Activity

▲ Match

Materials: Match Time Cards (BLM)

Lay cards in a facedown array. Have students turn over two cards, trying to match two Match Time Cards (BLM) that show the same time.

For example:

Exercise 1 • page 173

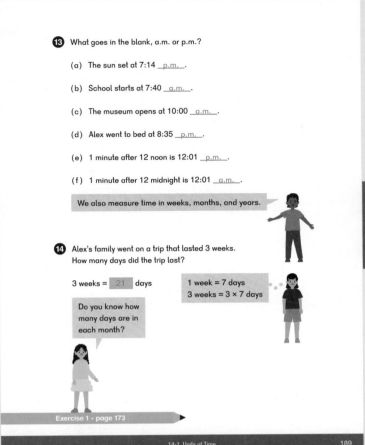

Lesson 2 Calculating Time — Part 1

Objectives

- Find elapsed time within a.m. or p.m.
- Find end time given start time and elapsed time, within a.m. or p.m.
- Add time in hours and minutes, using mental math strategies for "making 60."

Lesson Materials

- Student clocks with geared hands
- Demonstration clock

Think

Pose the **Think** problems and provide students with clocks to find the times.

Discuss student strategies for solving the problems. Use the demonstration clock to show students strategies.

Learn

In both (a) and (b), students make the next hour, a strategy similar to making the next ten. Students can use a number bond to see how many minutes are needed to get to the next full hour, and then how many minutes remain.

(a) 8:50 a.m. to 9:00 a.m. is 10 minutes.

40 minutes past 9:00 a.m. is 9:40 a.m.

```
      50 minutes
        /    \
10 minutes  40 minutes
```

(b) 9:40 a.m. to 10:00 a.m. is 20 minutes.

10:00 a.m. to 10:15 a.m. is 15 minutes.

20 minutes + 15 minutes = 35 minutes

(c) To find the elapsed time from the start of the Raptor Show until noon, add the hours first. This strategy is similar to "add the tens first" that students have used in prior lessons.

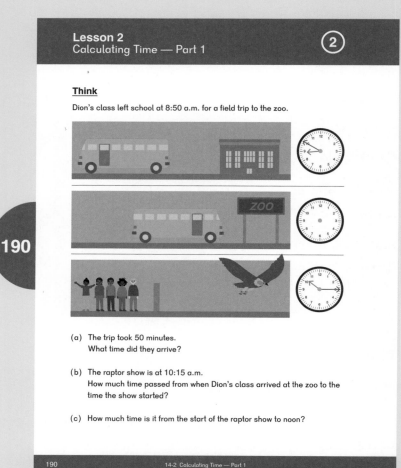

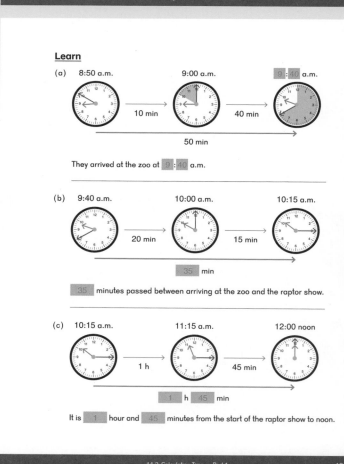

Do

1–**2** Students can use geared clocks to complete the problems.

1 (a) makes the next hour.
(b) passes through the next hour.
(c) passes through several hours.
(d) passes through hours and minutes.

3 Discuss the timeline example. Students should see that the timeline represents the times on a clock laid out in a line. Students can use this strategy instead of images of clocks.

4 Dion is using the strategy of making the next hour. He knows 60 minutes = 1 hour, so he makes 1 hour first, then adds the 5 minutes.

Emma adds the minutes and converts 60 minutes to one hour, so 65 minutes is 1 hour 5 minutes.

5 Sofia adds the hours first to get to 10:30 a.m. She then adds 45 minutes by adding 30 minutes to make the next hour, then adds 15 minutes.

6 Students could also add hours and minutes and convert, similar to Emma's strategy in **4**:

- 2 hours + 1 hour = 3 hours
- 45 minutes + 20 minutes = 65 minutes
- 3 hours 65 minutes = 4 hours 5 minutes

Exercise 2 • page 177

Do

1 How much time passes from...

(a) 3:25 p.m. to 4:00 p.m.?
   35 min
(b) 3:25 p.m. to 4:15 p.m.?
   50 min
(c) 3:25 p.m. to 7:25 p.m.?
   4 h
(d) 3:25 p.m. to 7:40 p.m.?
   4 h 15 min

2 What time is it...

(a) 8 minutes after 6:45 a.m.?
   6:53 a.m.
(b) 15 minutes after 6:45 a.m.?
   7:00 a.m.
(c) 35 minutes after 6:45 a.m.?
   7:20 a.m.
(d) 3 hours after 6:45 a.m.?
   9:45 a.m.
(e) 3 hours and 33 minutes after 6:45 a.m.?
   10:18 a.m.

3 A show started at 7:30 p.m. and ended at 10:15 p.m. How long was the show?
2 h 45 min

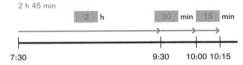

7:30   9:30   10:00 10:15

4 Alex went to bed, read for 50 minutes in bed, then turned off his light and fell asleep 15 minutes later. How long was it from when he went to bed until he fell asleep?   1 h 5 min

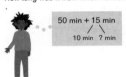 50 min + 15 min
10 h ? min

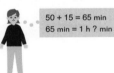 50 + 15 = 65 min
65 min = 1 h ? min

5 Mei went on a hike that took 2 hours and 45 minutes. The hike started at 8:30 a.m. When did it end?

8:30 $\xrightarrow{+2\,h}$ 10:30 $\xrightarrow{+45\,min}$ 11:15 a.m.

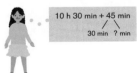 10 h 30 min + 45 min
30 min ? min

6 Dion spent 2 hours and 45 minutes reading. He spent another 1 hour and 20 minutes playing a game. How much time did he spend on both activities?

2 h 45 min $\xrightarrow{+1\,h}$ 3 h 45 min $\xrightarrow{+20\,min}$ 4 h 5 min

2 h 45 min + 1 h 20 min = 4 h 5 min

Exercise 2 • page 177

Lesson 3 Practice A

Objective

- Practice concepts from the first two lessons.

Lesson Materials

- Stopwatch, timer, or clock with a second hand

Have students complete the practice before moving on to the next lessons.

1 (b) Students may need help to see that there are six 10-second periods in one minute. Example:

- In 10 seconds, I can hop 7 times.
- In 20 seconds, I can hop 2 × 7 times.
- In 30 seconds, I can hop 3 × 7 times.
- …
- In 60 seconds, I can hop 6 × 7 times.

Exercise 3 • page 180

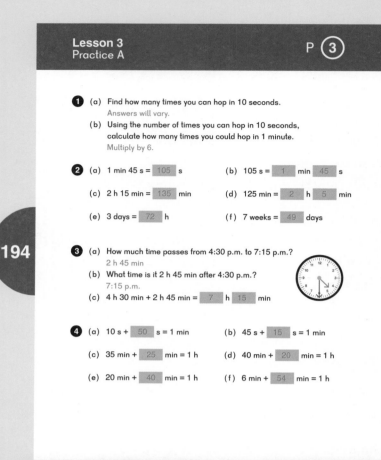

Lesson 4 Calculating Time — Part 2

Objectives

- Add periods of time in hours and minutes where the time changes from a.m. to p.m. or from p.m. to a.m.
- Find end times, when given start and elapsed times.

Lesson Materials

- Student clocks with geared hands
- Demonstration clock
- Timeline (BLM)

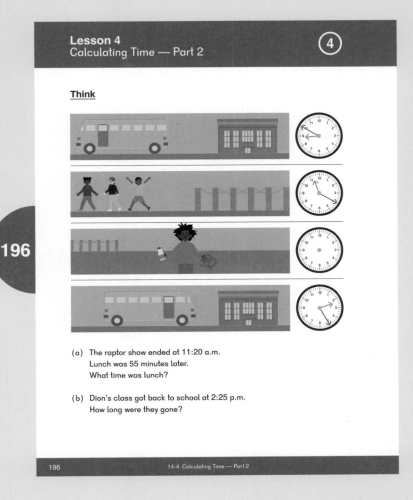

Think

Provide students with clocks and pose the **Think** problems. Have students try to solve them independently.

(a) Students may find the time that lunch starts by:

- drawing a timeline.
- thinking of 55 minutes as 40 minutes and 15 minutes to get to 12:00 p.m. noon, then see 15 minutes later.
- skip count by five or ten minute increments.

(b) Ensure students understand that the bus leaves at 8:50 a.m. and returns to school at 2:25 p.m.

Discuss students' strategies for solving the problems.

Learn

Have students compare their solutions from **Think** with those shown in the textbook.

(a) Students think, "From 11:20 a.m. I need to add 55 minutes. How many minutes until 12:00 noon? How many minutes past noon?"

Note that students may use a mental math strategy similar to over-adding: 55 minutes is the same as adding 1 hour then subtracting or counting back 5 minutes.

For example:

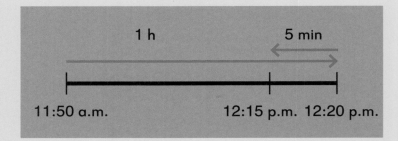

(b) **Method 1** counts on the whole hours, then adds the minutes.

Students count on hours:

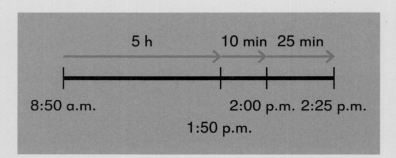

Method 2 adds hours and minutes to get to 12:00 noon. Having reached noon, students can next count hours and minutes again, starting with 1.

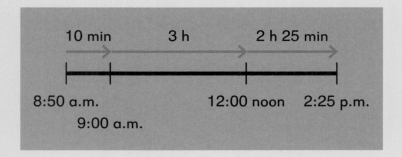

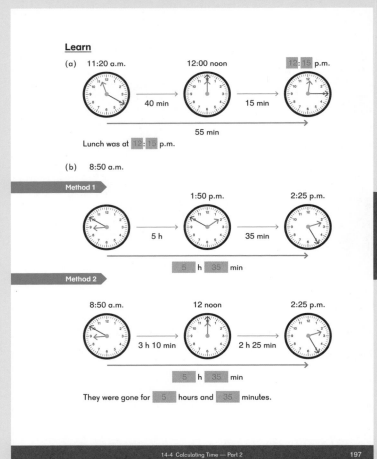

This strategy is provided so that students do not get confused as the hour numbers go back to 1. With practice, some students may not need to use this strategy.

Provide students with the Timeline (BLM) to practice as needed.

Students may find other methods that work for them. The focus should be on finding an efficient method.

Do

4 – 8 Students who are struggling can use student clocks, draw timelines, or use a Timeline (BLM) if needed. Most students should be working from the textbook.

Have students share how they found their answers.

5 Mei uses **Learn** Method 1, and first adds the hours, then the minutes.

Students may count on hours from 8:15 a.m. to 3:15 p.m., or they may know 8:15 a.m. to 12:15 p.m. is 4 hours and 12:15 p.m. to 3:15 p.m. is 3 hours.

4 + 3 = 7 hours

Emma uses **Learn** Method 2. She adds 3 h 45 min to get to 12:00 noon, and then adds 3 h 50 min to get her answer.

7 This is the first problem that finds time to the 1 minute.

Exercise 4 • page 183

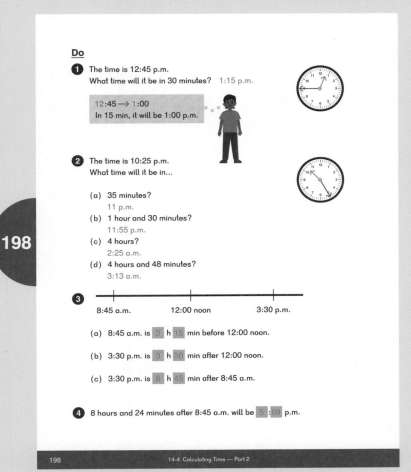

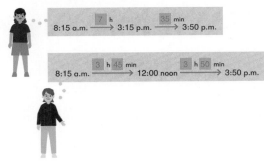

Lesson 5 Calculating Time — Part 3

Objectives

- Subtract periods of time in hours and minutes that do not cross a.m. or p.m.
- Find start times, when given the end times and elapsed time.

Lesson Materials

- Student clocks with geared hands
- Timeline (BLM)

Think

Provide students with geared clocks or Timeline (BLM) and pose the **Think** problems. Have students try to solve them independently.

Discuss students' strategies for solving the problem.

Note that students are comparing the friends' times:

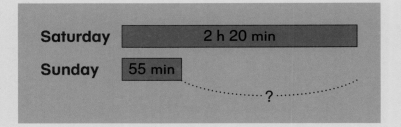

Learn

Have students compare their solutions from **Think** with Sofia, Dion, and Emma's methods.

Students may find this material challenging, especially because they are still learning to regroup to make 60.

Method 1

Sofia subtracts 55 minutes from 1 hour to get a difference of 5 minutes. She then adds 1 h 20 min and 5 min.

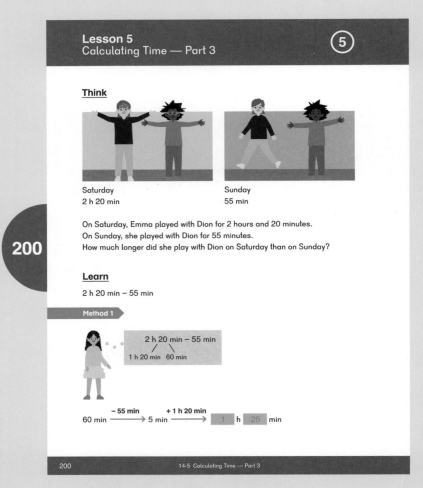

Method 2

Dion uses the over-subtracting strategy. 60 minutes is an hour, so he subtracts an hour and adds back 5 minutes.

Method 3

Emma splits the 55 minutes into 20 minutes and 35 minutes, and subtracts twice. First she subtracts 20 minutes from 2 h 20 min to leave 2 h. She then subtracts 35 minutes from 2 hours.

Students may find other methods that work for them. The focus is on finding an efficient method.

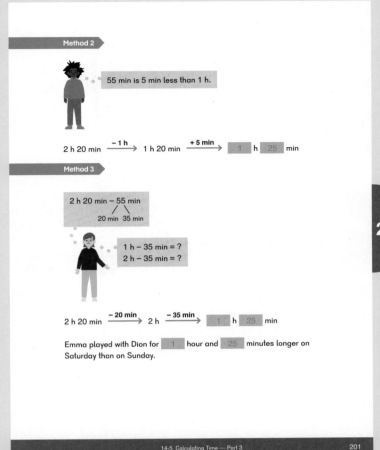

Do

Students who are struggling can use student clocks, draw timelines, or use a Timeline (BLM) if needed. Most students should be working from the textbook.

Have students share how they found their answers.

④ Alex first subtracts the hours, then the minutes.

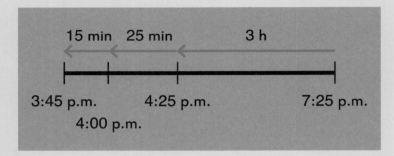

Mei subtracts 25 minutes to get to the hour, 7 p.m., then subtracts another 3 hours 15 minutes.

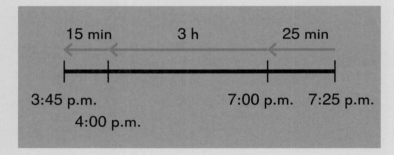

Exercise 5 • page 187

Do

① What time is...
  (a) 6 minutes before 6:15 p.m.? 6:09 p.m.
  (b) 15 minutes before 6:15 p.m.? 6:00 p.m.
  (c) 35 minutes before 6:15 p.m.? 5:40 p.m.
  (d) 3 hours before 6:15 p.m.? 3:15 p.m.
  (e) 3 hours and 30 minutes before 6:15 p.m.? 2:45 p.m.

② (a) 1 h – 25 min = 35 min
  (b) 3 h – 25 min = 2 h 35 min
  (c) 3 h 10 min – 25 min = 2 h 45 min
  (d) 3 h 50 min – 26 min = 3 h 24 min

③ Subtract.
  (a) 1 h – 20 min | 2 h – 20 min | 2 h 10 min – 20 min
     40 min | 1 h 40 min | 1 h 50 min
  (b) 1 h – 45 min | 6 h – 45 min | 6 h 50 min – 45 min
     15 min | 5 h 15 min | 6 h 5 min
  (c) 1 h – 3 min | 3 h – 3 min | 3 h 50 min – 3 min
     57 min | 2 h 57 min | 3 h 47 min

④ What time is 3 h 40 min before 7:25 p.m.?

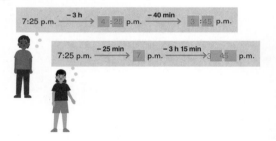

⑤ (a) 5 h 15 min – 45 min = 4 h 30 min
  (b) 5 h 15 min – 2 h = 3 h 15 min
  (c) 5 h 15 min – 2 h 45 min = 2 h 30 min

⑥ Subtract.
  (a) 1 h 55 min – 1 h 35 min   (b) 3 h 5 min – 1 h 45 min
     20 min   1 h 20 min
  (c) 6 h 10 min – 1 h 30 min   (d) 5 h 20 min – 1 h 32 min
     4 h 40 min   3 h 48 min

⑦ Sofia raked her yard for 1 hour and 50 minutes. She finished at 4:25 p.m. What time did she start raking her yard?
  2:35 p.m.

Exercise 5 • page 187

Lesson 6 Calculating Time — Part 4

Objective

- Subtract periods of time in hours and minutes that cross a.m. and p.m.

Lesson Materials

- Student clocks with geared hands
- Timeline (BLM)

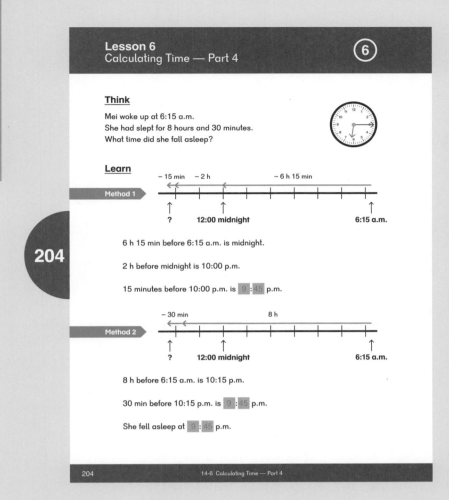

Think

Provide students with clocks and pose the **Think** problem. Have students try to solve the problem independently. They should see that they are looking for an earlier time.

Discuss students' strategies for solving the problem.

Learn

Have students compare their solutions from **Think** with the ones shown in the textbook.

Method 1

On the timeline, 6 hours 15 minutes are subtracted to get to 12:00 midnight. From there, the remaining hours are subtracted to get to 10:00 p.m., and then the remaining minutes to get to 9:45 p.m.

Method 2

On the timeline, the hours are first subtracted to get to 10:15 p.m., then the remaining minutes to get to 9:45 p.m.

© 2017 Singapore Math Inc.   Teacher's Guide 3B Chapter 14   227

Do

1 – 2 Students who are struggling can use student clocks, draw timelines, or use a Timeline (BLM) if needed. Most students should be working from the textbook.

5 To find the time Siti should leave her home, students should count back 3 h 30 min from 2:30 p.m.

Exercise 6 • page 191

Do

1 The time is 1:05 p.m.
What time was it...

(a) 35 minutes earlier?
  12:30 p.m.
(b) 1 hour and 5 minutes earlier?
  12:00 p.m.
(c) 1 hour and 35 minutes earlier?
  11:30 a.m.
(d) 4 hours earlier?
  9:05 a.m.
(e) 4 hours and 45 minutes earlier?
  8:20 a.m.

2 The time is 5:30 p.m.
What time was it...

(a) 5 hours and 30 minutes earlier?
  12:00 p.m.
(b) 6 hours earlier?
  11:30 a.m.
(c) 6 hours and 15 minutes earlier?
  11:15 a.m.
(d) 12 hours earlier?
  5:30 a.m.
(e) 10 hours earlier?
  7:30 a.m.
(f) 10 hours and 45 minutes earlier?
  6:45 a.m.

3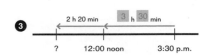

?   12:00 noon   3:30 p.m.

(a) 12 noon is 3 hours and 30 minutes before 3:30 p.m.
(b) 2 hours and 20 minutes before 12 noon is 9:40 a.m.
(c) What time is 5 hours and 50 minutes before 3:30 p.m.?
  9:40 a.m.

4 (a) 20 min before midnight is 11:40 p.m.
(b) 2 h 20 min before midnight is 9:40 p.m.
(c) 2 h 20 min before 1 a.m. is 10:40 p.m.
(d) 4 h before 2:30 p.m. is 10:30 a.m.
(e) 4 h 50 min before 2:30 p.m. is 9:40 a.m.

5 The ferry leaves at 2:30 p.m.
Cars should be in line 1 hour before the departure time.
It takes 2 hours 30 minutes to get from Siti's home to the ferry.
What time should she leave home?
11:00 a.m.

Exercise 6 • page 191

Lesson 7 Practice B

Objective
- Practice concepts from the chapter.

After students complete **Practice B**, have them continue finding elapsed time.

3 — **5** Have students share how they subtracted the units of time for some of the problems.

Exercise 7 • page 195

Brain Works

★ Time Puzzles

Airplanes take off from a runway every 30 seconds. How many flights leave in 5 minutes? 10 minutes? 1 hour?

1 minute ⟶ 2 flights
5 minutes ⟶ 5 × 2 flights = 10 flights
10 minutes ⟶ 10 × 2 flights = 20 flights
60 minutes ⟶ 60 × 2 flights = 120 flights

Trains leave a station every 16 minutes. If the first train leaves at 6:00 a.m., when does the 6th train leave?

6 × 16 = 96 minutes = 1 h 36 min
1 h 36 min after 6:00 a.m. is 7:36 a.m.

Lesson 7 Practice B   P 7

1 Subtract.
(a) 55 min − 35 min
    20 min
(b) 12 h − 7 h
    5 h
(c) 1 h − 35 min
    25 min
(d) 5 h − 35 min
    4 h 25 min
(e) 5 h − 3 h 35 min
    1 h 25 min
(f) 11 h − 2 h 12 min
    8 h 48 min

2 (a) This clock is 6 minutes fast. What is the correct time?
    10:09
(b) This clock is 6 minutes slow. What is the correct time?
    3:03

3 Subtract.
(a) 5 h − 45 min
    4 h 15 min
(b) 5 h 10 min − 45 min
    4 h 25 min
(c) 5 h 10 min − 2 h 45 min
    2 h 25 min
(d) 5 h 55 min − 2 h 45 min
    3 h 10 min
(e) 12 h − 20 min
    11 h 40 min
(f) 12 h 40 min − 20 min
    12 h 20 min
(g) 12 h 40 min − 50 min
    11 h 50 min
(h) 12 h 40 min − 6 h 50 min
    5 h 50 min
(i) 4 h 10 min − 2 h 45 min
    1 h 25 min
(j) 3 h 15 min − 2 h 50 min
    25 min

4 What time is 3 hours 50 minutes before 1:45 p.m.?
9:55 a.m.

5 Lucia had a 45-minute piano lesson. The lesson was over at 10:15 a.m. What time did her lesson start?
9:30 a.m.

6 A baseball game began at 12:25 p.m. It lasted 2 hours and 28 minutes. What time did the game end?
2:53 p.m.

7 Andrei arrived at the amusement park at 9:50 a.m. He left the amusement park at 4:45 p.m. How long was he there?
6 h 55 min

8 It took Larry 2 hours and 35 minutes to paint the living room. It took him another 1 hour and 55 minutes to paint the bedroom.
(a) How long did it take him to paint both rooms?
    4 h 30 min
(b) How much longer did it take him to paint the living room than the bedroom?
    40 min

9 A school play ended at 1:15 p.m. The play lasted 2 hours and 30 minutes. What time did the play start?
10:45 a.m.

Exercise 7 • page 195

Exercise 1 • pages 173–176

Chapter 14 Time

Exercise 1

Basics

1. Fill in the blanks with min for minutes, s for seconds, or h for hours.

   (a) You should wash your hands for at least 20 __s__.

   (b) You should brush your teeth for at least 2 __min__.

   (c) Colton's piano recital lasted for 2 __h__.

   (d) Aliya ran halfway around the football field in about 45 __s__.

   (e) A soccer game has two 45-__min__ halves.

   (f) It takes about 6 __h__ to fly by airplane across the United States.

2. Write a.m. or p.m. in the blanks.

   (a) 10:00 __a.m.__ is between midnight and noon.

   (b) Jamal went to bed for the night at 9:00 __p.m.__

   (c) 5 hours after 3:00 a.m. is 8:00 __a.m.__

   (d) 4 hours before 3:00 a.m. is 11:00 __p.m.__

   (e) The restaurant is open from 6:30 _____ to 10:00 _____ .p.m.

3. (a) 1 h = 60 min
   3 h = __180__ min
   3 h 15 min = __195__ min
   (b) 60 min = __1__ h
   120 min = __2__ h
   130 min = __2__ h __10__ min

4. (a) 1 min = 60 s
   2 min = __120__ s
   2 min 45 s = __165__ s
   (b) 60 s = __1__ min
   180 s = __3__ min
   200 s = __3__ min __20__ s

5. (a) 1 day = __24__ h
   5 days = __120__ h
   (b) 1 week = __7__ days
   15 weeks = __105__ days

Practice

6. Complete the tables.

Hours	Minutes
1	60
2	120
3	180
4	240
5	300

Hours	Minutes
6	360
7	420
8	480
9	540
10	600

7. Match.

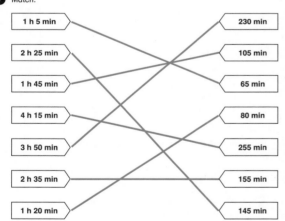

8. (a) 5 min 5 s = __305__ s
   (b) 1 min 32 s = __92__ s
   (c) 140 s = __2__ min __20__ s
   (d) 335 s = __5__ min __35__ s

9. (a) 1 week 2 days = __9__ days
   (b) 16 days = __2__ weeks __2__ days

10. Dexter went away on a trip that lasted 3 weeks and 4 days. How many days was he gone?
    $3 \times 7 + 4 = 25$
    He was gone 25 days.

11. Valentina practiced the bass for 85 minutes. Onowa practiced the piano for 1 hour 20 minutes. Who practiced longer and how much longer?
    85 min = 1 h 25 min or 1 h 20 min = 80 min
    Valentina practiced 5 minutes longer.

12. Caden ran a marathon in 228 minutes. Diego ran the same marathon in 4 hours 13 minutes. Who ran the marathon faster and how much faster?
    4×60 min + 13 min = 253 min
    4 h 13 min = 253 min
    253 min − 228 min = 25 min
    Caden ran the marathon faster by 25 min.

Challenge

13. 366 days = __52__ weeks __2__ days
    366 ÷ 7 is 52 with a remainder of 2.

14. How many days are in a month that begins and ends on a Wednesday?
    4 weeks and 1 day
    29 days

Exercise 2 • pages 177–179

Exercise 2

Basics

1. Complete the following:

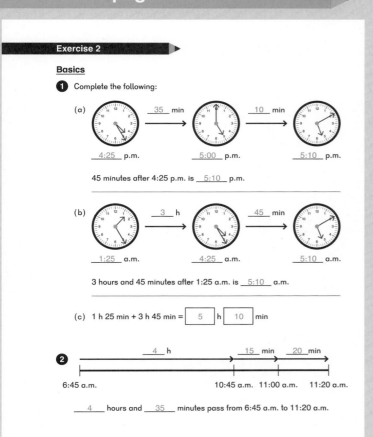

(a) 45 minutes after 4:25 p.m. is 5:10 p.m.

(b) 3 hours and 45 minutes after 1:25 a.m. is 5:10 a.m.

(c) 1 h 25 min + 3 h 45 min = 5 h 10 min

2.   4 hours and 35 minutes pass from 6:45 a.m. to 11:20 a.m.

Practice

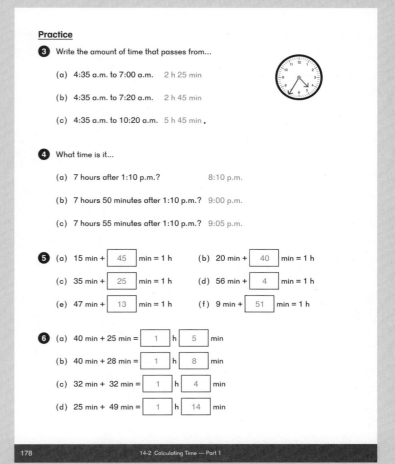

3. Write the amount of time that passes from…

(a) 4:35 a.m. to 7:00 a.m.   2 h 25 min

(b) 4:35 a.m. to 7:20 a.m.   2 h 45 min

(c) 4:35 a.m. to 10:20 a.m.   5 h 45 min.

4. What time is it…

(a) 7 hours after 1:10 p.m.?   8:10 p.m.

(b) 7 hours 50 minutes after 1:10 p.m.?   9:00 p.m.

(c) 7 hours 55 minutes after 1:10 p.m.?   9:05 p.m.

5. (a) 15 min + 45 min = 1 h   (b) 20 min + 40 min = 1 h
   (c) 35 min + 25 min = 1 h   (d) 56 min + 4 min = 1 h
   (e) 47 min + 13 min = 1 h   (f) 9 min + 51 min = 1 h

6. (a) 40 min + 25 min = 1 h 5 min
   (b) 40 min + 28 min = 1 h 8 min
   (c) 32 min + 32 min = 1 h 4 min
   (d) 25 min + 49 min = 1 h 14 min

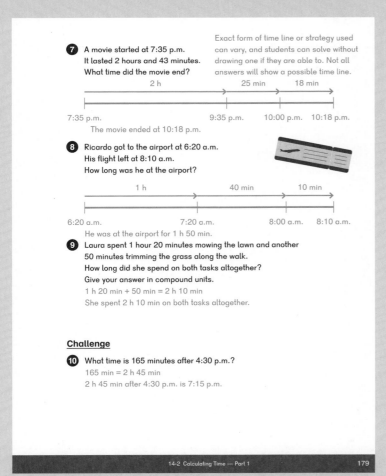

7. A movie started at 7:35 p.m. It lasted 2 hours and 43 minutes. What time did the movie end?

Exact form of time line or strategy used can vary, and students can solve without drawing one if they are able to. Not all answers will show a possible time line.

7:35 p.m. → 2 h → 9:35 p.m. → 25 min → 10:00 p.m. → 18 min → 10:18 p.m.

The movie ended at 10:18 p.m.

8. Ricardo got to the airport at 6:20 a.m. His flight left at 8:10 a.m. How long was he at the airport?

6:20 a.m. → 1 h → 7:20 a.m. → 40 min → 8:00 a.m. → 10 min → 8:10 a.m.

He was at the airport for 1 h 50 min.

9. Laura spent 1 hour 20 minutes mowing the lawn and another 50 minutes trimming the grass along the walk. How long did she spend on both tasks altogether? Give your answer in compound units.

1 h 20 min + 50 min = 2 h 10 min
She spent 2 h 10 min on both tasks altogether.

Challenge

10. What time is 165 minutes after 4:30 p.m.?
    165 min = 2 h 45 min
    2 h 45 min after 4:30 p.m. is 7:15 p.m.

Exercise 3 • pages 180–182

Exercise 3

Check

1. This chart shows the time it took for some girls to swim 200 meters.

Renata	165 s
Taylor	2 min 15 s
Natalia	190 s
Violet	3 min 5 s

   Write the times in order from shortest time to longest time.
   2 min 15 s, 165 s, 3 min 5 s, 190 s

2. (a) 3 h 40 min + 4 h 50 min = **8** h **30** min
   (b) 5 h 50 min + 2 h 43 min = **8** h **33** min
   (c) 11 h 17 min + 6 h 25 min = **17** h **42** min
   (d) 1 h 10 min + 2 h 25 min + 3 h 45 min = **7** h **20** min

3. Write the amount of time that passes from...
   (a) 2:35 p.m. to 4:00 p.m.   1 h 25 min
   (b) 8:50 a.m. to 11:39 a.m.   2 h 49 min
   (c) 1:22 p.m. to 3:40 p.m.   2 h 18 min

4. If yesterday was Tuesday, what day of the week is it 1 week and 4 days from today?
   Today is Wednesday, 1 week from today is Wednesday, 4 more days is Sunday.
   Sunday.

5. Kim took 62 days to hike the Pacific Northwest Trail. Write this time in weeks and days.
   62 ÷ 7 is 8 with a remainder of 6.
   8 weeks 6 days

6. A show began at 4:35 p.m. and ended at 6:20 p.m. How long did the show last?

   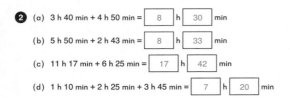

   The show lasted 1 h 45 min.

7. Megan started exercising at 6:12 a.m. She exercised for 1 hour 42 minutes. What time did she finish exercising?

   6:12 a.m. — 1 h — 7:12 a.m. — 42 min — 7:54 a.m.

   She finished exercising at 7:54 a.m.

8. Isaac left home at 8:50 a.m.
   He took 33 minutes to travel to the park.
   He spent 2 hours 20 minutes at the park.
   What time did he leave the park?
   8 h 50 min + 33 min + 2 h 20 min = 11 h 43 min
   He left the park at 11:43 a.m.

9. Heather started an art project at 4:30 p.m.
   It took her 15 minutes to gather the material and set up her canvas.
   She spent 2 hours 20 minutes painting.
   She spent another 25 minutes cleaning up.
   What time did she finish?
   15 min + 2 h 20 min + 25 min = 3 h
   She finished at 7:30 p.m.

Challenge

10. A bus service runs every 15 minutes.
    There is a bus at 9:10 a.m.
    How many buses are there between 9:00 a.m. and 11:00 a.m.?
    9:10, 9:25, 9:40, 9:55, 10:10, 10:25, 10:40, 10:55
    There are 8 buses.

Exercise 4 • pages 183–186

Exercise 4

Basics

1. Write the times using a.m. or p.m.

   (a) 8:50 a.m. → 4 hours later → 12:50 p.m.

   (b) 8:50 a.m. → 4 hours 35 min later → 1:25 p.m.

   (c) 6 hours and 35 minutes after 8:50 a.m. is __3:25 p.m.__.

2. (a) 8:45 p.m. to midnight is __3__ hours and __15__ minutes.

   (b) Midnight to 6:30 a.m. is __6__ hours and __30__ minutes.

   (c) 8:45 p.m. to 6:30 a.m. is __9__ hours and __45__ minutes.

3. Complete the following:

   7 h — 25 min
   10:50 a.m. | 12:00 noon | __5:50__ p.m. | __6:15__ p.m.

   7 h 25 minutes after 10:50 a.m. is __6:15 p.m.__.

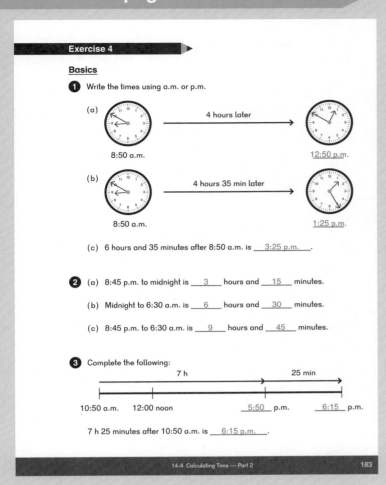

Practice

4. Write the amount of time that passes from...

   (a) 10:40 a.m. to 12:00 noon. 1 h 20 min
   (b) 10:40 a.m. to 1:40 p.m. 3 h
   (c) 10:40 a.m. to 2:15 p.m. 3 h 35 min
   (d) 10:40 a.m. to 2:18 p.m. 3 h 38 min
   (e) 10:42 a.m. to 2:15 p.m. 3 h 33 min
   (f) 10:42 a.m. to 2:18 p.m. 3 h 36 min

5. What time is it...

   (a) 8 hours 50 minutes after 7:25 p.m.? 4:15 a.m.

   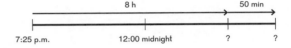

   7:25 p.m. | 12:00 midnight | ? | ?
   8 h | 50 min

   (b) 8 hours 55 minutes after 7:25 p.m.? 4:20 a.m.
   (c) 8 hours 57 minutes after 7:25 p.m.? 4:22 a.m.
   (d) 8 hours 55 minutes after 7:29 p.m.? 4:24 a.m.
   (e) 8 hours 57 minutes after 7:29 p.m.? 4:26 a.m.

Methods may vary.

6. A store opens at 6:30 a.m. and closes at 10:45 p.m.

   (a) How long is it open during the day?
   16 h 15 min

   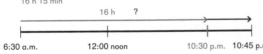
   6:30 a.m. | 12:00 noon | 10:30 p.m. | 10:45 p.m.
   16 h | ?

   (b) How long is it closed during the night?
   24 h − 16 h 15 min = 7 h 45 min
   7 h 45 min

7. Mr. Ikeda left home for work at 7:35 a.m.
   He arrived back home 9 hours and 45 minutes later.
   What time did he arrive back home? He arrived back home at 5:20 p.m.

   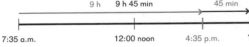
   7:35 a.m. | 12:00 noon | 4:35 p.m. | ?
   9 h | 9 h 45 min | 45 min

8. Sasha went to bed at 8:50 p.m.
   She read for 25 minutes and fell asleep 13 minutes after she finished reading.
   She slept for 8 hours and 10 minutes.
   What time did she wake up?
   8:50 p.m. + 38 min + 8 h 10 min = 5:38 a.m.
   She woke up at 5:38 a.m.

9. Laura started working in the yard at 11:25 a.m.
   She weeded her garden for 1 hour and 30 minutes.
   She trimmed her fruit trees for 2 hours and 15 minutes.
   Then she spent 25 minutes tidying up and putting away her tools.
   At what time did she finish?
   1 h 30 min + 2 h 15 min + 25 min = 4 h 10 min
   4 h 10 min after 11:25 a.m. is 3:35 p.m.
   She finished at 3:35 p.m.

Challenge

10. How many hours and minutes are there between 9:45 p.m. on Friday and 6:15 a.m. on Sunday?
    9:45 p.m. Friday to 6:15 a.m. Saturday is 8 h 30 min.
    8 h 30 min + 24 h = 32 h 30 min
    32 h 30 min

11. A bus service runs every 25 minutes.
    There is a bus at 9:10 a.m.
    How many buses are there between 9:00 a.m. and 4:00 p.m.?
    9:10, 9:35, 10:00, 10:25, 10:50, 11:15, 11:40, 12:05,
    12:30, 12:55, 1:20, 1:45, 2:10, (then pattern repeats)
    2:35, 3:00, 3:25, 3:50
    There are 17 buses.

© 2017 Singapore Math Inc.   Teacher's Guide 3B Chapter 14

Exercise 5 • pages 187–190

Exercise 5

Basics

1 (a) 3 h − 45 min = 2 h [15] min
    2 h   60 min

(b) 3 h 10 min − 45 min = 2 h [25] min
    2 h 10 min   60 min

(c) 7 h 10 min —−4 h→ 3 h 10 min —−45 min→ [2] h [25] min

(d) 4 hours 45 minutes before 7:10 a.m. is __2:25 a.m.__ .

2 (a) 3 h − 10 min = 2 h [50] min

(b) 4 h 25 min − 35 min = [3] h [50] min
    25 min 10 min

(c) 4 h 25 min − 1 h 35 min = [2] h [50] min

(d) 1 hour 35 minutes before 4:25 p.m. is __2:50 p.m.__ .

3 (a) 54 min = 1 h − [6] min

(b) 3 h 20 min —−1 h→ 2 h 20 min —+6 min→ [2] h [26] min

(c) 5 h 20 min − 2 h 54 min = [2] h [26] min

(d) 2 hours 54 minutes before 5:20 p.m. is __2:26 p.m.__ .

Practice

4 What time is it...

(a) 35 minutes before 4:35 p.m.? 4:00 p.m.

(b) 45 minutes before 4:35 p.m.? 3:50 p.m.

(c) 1 hour and 45 minutes before 4:35 p.m.? 2:50 p.m.

(d) 1 hour and 48 minutes before 4:35 p.m.? 2:47 p.m.

5 (a) 5 h − 32 min = [4] h [28] min

(b) 5 h 10 min − 32 min = [4] h [38] min

6 (a) 5 h 55 min − 2 h 24 min = [3] h [31] min

(b) 7 h 32 min − 3 h 50 min = [3] h [42] min

(c) 20 h 5 min − 18 h 30 min = [1] h [35] min

7 What time is it...

(a) 8 hours 25 minutes before 10:05 p.m.?
    1:40 p.m.

(b) 3 hours 55 minutes before 11:35 a.m.?
    7:40 a.m.

Methods may vary.

8 Mia spent 3 hours 45 minutes at a park.
She left the park at 4:35 p.m.
What time did she get to the park?

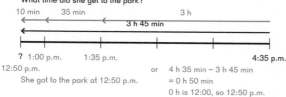

? 1:00 p.m. 1:35 p.m.   4:35 p.m.
12:50 p.m.   or   4 h 35 min − 3 h 45 min
She got to the park at 12:50 p.m.   = 0 h 50 min
                     0 h is 12:00, so 12:50 p.m.

9 Megan exercised for 1 hour 25 minutes on Monday.
She exercised for 2 hours 15 minutes on Wednesday.
How much longer did she exercise on Wednesday than on Monday?
2 h 15 min − 1 h 25 min = 50 min
or: 35 min from 1 h 25 min to 2 h, plus another 15 min is 50 min
She exercised 50 min longer on Wed. than on Mon.

10 Arman spent 5 hours 15 minutes at the lake.
He spent 35 minutes having a picnic, 1 hour 20 minutes swimming,
and the rest of the time fishing.
How much time did he spend fishing?
1 h 20 min + 35 min = 1 h 55 min
5 h 15 min − 1 h 55 min = 3 h 20 min
or: 1 h 55 min to 2 h is 5 min, then to 5 h is 3 h 5 min,
then another 15 min is 3 h 20 min.
He spent 3 h 20 min fishing.

11 A concert will start at 7:30 p.m.
Ethan wants to reach the performance hall 15 minutes before the start time.
It will take him 1 hour 25 minutes to get to the hall from his house.
What time should he leave his house?

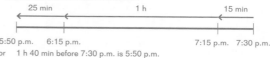

     25 min   1 h   15 min
5:50 p.m. 6:15 p.m.   7:15 p.m. 7:30 p.m.
or   1 h 40 min before 7:30 p.m. is 5:50 p.m.
     He should leave his house at 5:50 p.m.

Challenge

12 A clock loses 3 minutes every hour.
If it was set to the correct time at 10:00 a.m. on Friday, what time will it show at 10:00 a.m. on Saturday?
24 × 3 min = 72 min
It loses 72 min, or 1 h 12 min.
1 h 12 min before 10:00 a.m. is 8:48 a.m.
It will show 8:48 a.m.

13 A bus service runs every 20 minutes.
There is a bus at 9:10 a.m.
How many buses are there between 6:00 a.m. and noon?
9:10 back: 8:50, 8:30, 8:10 (pattern repeats)
7:50, 7:30, 7:10, 6:50, 6:30, 6:10
9:10 on: 9:30, 9:50, 10:10 (pattern repeats)
10:30, 10:50, 11:10, 11:30, 11:50
(Remember to include 9:10 in the count.)
18 buses

Exercise 6 • pages 191–194

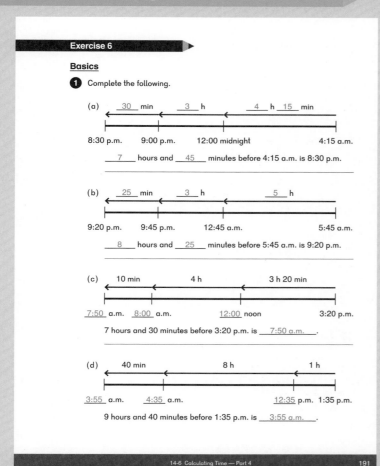

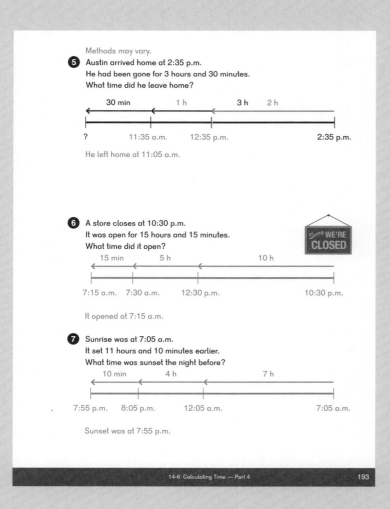

Exercise 7 • pages 195–198

Exercise 7

Check

1. (a) 7 h 10 min − 5 h 25 min = **1** h **45** min
   (b) 3 h 55 min − 2 h 24 min = **1** h **31** min
   (c) 3 h 25 min + **4** h **10** min = 7 h 35 min
   (d) 7 h 50 min + **2** h **25** min = 10 h 15 min
   (e) **1** h **37** min + 12 h 30 min = 14 h 7 min

2. Sunset was at 8:45 p.m.
   The sun rose 8 hours and 45 minutes later.
   What time was sunrise?
   8 hours after 8:45 p.m. is 4:45 a.m., 45 minutes after that is 5:30 a.m.
   Sunrise was at 5:30 a.m.

3. Papina practiced the violin for 1 hour 20 minutes.
   She then played outside with her friends for 3 hours 45 minutes.
   How much more time did she spend playing than practicing?
   3 h 45 min − 1 h 20 min = 2 h 25 min
   or: 2 h more than 1 h 20 min is 3 h 20 min, then another 25 min is 3 h 45 min.
   She spent 2 h 25 min more playing than practicing.

Methods may vary.

4. Colton started playing tennis at 11:45 a.m. and finished at 1:15 p.m.
   How many minutes did he play tennis?

   1 h 30 min = 90 min   He played for 90 min.

5. A program ended at 12:45 p.m.
   It was 2 hours and 30 minutes long.
   When did it start?

   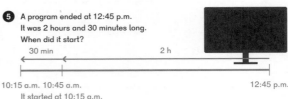

   It started at 10:15 a.m.

6. A restaurant opens every day at 11:00 a.m. for lunch, closes at 2:45 p.m., opens again at 5:30 p.m. for dinner, and closes at 11:30 p.m.
   How long is it open in one day?
   11:00 a.m. to 2:45 p.m. is 3 h 45 min.
   5:30 p.m. to 11:30 p.m. is 6 h.
   3 h 45 min + 6 h = 9 h 45 min
   It is open for 9 h 45 min.

7. Misha did a project on millipedes for school.
   She spent 1 hour 35 minutes researching the topic and another 45 minutes writing the report.
   She finished her report at 5:10 p.m.
   What time did she start her project?
   1 h 35 min + 45 min = 2 h 20 min
   5 h 10 min − 2 h 20 min = 2 h 50 min
   She started at 2:50 p.m.

Challenge

8. What time will it be 800 minutes after 10:30 a.m.?
   600 minutes is 10 hours
   200 more minutes is another 3 hours and 20 minutes.
   13 hours and 20 minutes after 10:30 a.m. is 11:50 p.m.

9. Sydney has an analog clock and a digital clock that both run on electricity. When the power goes out, both stop running, but only the digital clock resets to 12:00 midnight.

   (a) One night, the power went off at 3:00 a.m. for 25 minutes.
       At 6:15 a.m., what time will show on each clock?
       Analog: It loses 25 minutes while the power is off.
       25 minutes before 6:15 a.m. is 5:50 a.m.
       Digital: At 3:00 a.m., the clock resets to 12:00 midnight and stays there for another 25 minutes, so it loses 3 hours and 25 minutes.
       3 hours and 25 minutes before 6:15 a.m. is 2:50 a.m.

   (b) Another night, Sydney woke up to find her analog clock displaying 5:05 and her digital clock displaying 2:45 a.m.
       What time was it when the power went out?
       From (a), students should see that the difference in time between the two clocks, 3 hours, is the time the power went off.
       The difference in time from 2:45 to 5:05 is 2 hours and 20 minutes.
       The power went off at 2:20 a.m.

10. There are 365 days in a year, except for a leap year, which has 366 days.
    The year 2000 was a leap year, and every 4 years after that.
    The chart shows the number of days in each month in a non-leap year.

Jan.	Feb.	Mar.	Apr.	May	June	July	Aug.	Sept.	Oct.	Nov.	Dec.
31	28	31	30	31	30	31	31	30	31	30	31

    Every leap year, February has 29 days instead of 28.

    Write down your birth date and today's date.
    How many days old are you?
    Answers will vary.

Chapter 15 Money

Overview

Suggested number of class periods: 8–9

	Lesson	Page	Resources		Objectives
	Chapter Opener	p. 241	TB:	p. 209	Investigate making change.
1	Dollars and Cents	p. 242	TB: WB:	p. 210 p. 199	Count and convert amounts of money within $100 from dollars and cents to cents, and from cents to dollars and cents.
2	Making $10	p. 245	TB: WB:	p. 214 p. 202	Add amounts that make $10. Subtract amounts from $10.
3	Adding Money	p. 247	TB: WB:	p. 216 p. 204	Add money amounts within $100.
4	Subtracting Money	p. 250	TB: WB:	p. 220 p. 207	Subtract money amounts within $100.
5	Word Problems	p. 252	TB: WB:	p. 224 p. 210	Solve word problems involving adding and subtracting money.
6	Practice	p. 254	TB: WB:	p. 227 p. 213	Practice skills from the chapter.
	Review 4	p. 256	TB: WB:	p. 229 p. 217	Cumulative review of content from Chapter 1 through Chapter 15.
	Review 5	p. 258	TB: WB:	p. 233 p. 222	Cumulative review of content from Chapter 1 through Chapter 15.
	Workbook Solutions	p. 260			

Chapter 15 Money

Notes

In Dimensions Math 2B, students learned to:

- Identify pennies, nickels, dimes, and quarters.
- Count change and bills.
- Add and subtract money within $20 using a mental math strategy.
- Convert dollars and cents to cents.
- Convert cents to dollars and cents, up to $10.

In this chapter, students will make change for amounts within $100. It is important that students are able to convert between amounts written in cents and dollars and cents:

$50.94 = 5,094¢
43¢ = $0.43

In both Dimensions Math 2A and 3A, students learned to add and subtract numbers mentally to make 100 or 1,000. In this chapter, they will be adding and subtracting to $10.

Example: $7.95 + ▇ = $10.00

Lessons 3 and 4 will apply these skills when adding and subtracting amounts of money.

Mental Math

Mental math methods introduced in Dimensions Math 3A Chapter 2: Addition and Subtraction — Part 1 will be practiced and applied in this chapter.

Addition Strategies

Add $17.80 + $9.60.

Add the dollars, then the cents:

$$\$17.80 \xrightarrow{+\$9} \$26.80 \xrightarrow{+60¢} \$27.40$$

Make the next dollar:

$$\$17.80 \xrightarrow{+20¢} \$18.00 \xrightarrow{+\$9.40} \$27.40$$

Add too many and subtract the difference, which can be called "over-adding."

$$\$17.80 \xrightarrow{+\$10} \$27.80 \xrightarrow{-40¢} \$27.40$$

Subtraction Strategies

Subtract $7.65 from $19.60.

Subtract the dollars, then the cents:

$$\$19.60 \xrightarrow{-\$7} \$12.60 \xrightarrow{-65¢} \$11.95$$

Make an even dollar:

$$\$19.60 \xrightarrow{-60¢} \$19.00 \xrightarrow{-\$7.05} \$11.95$$

Subtract too many and add the difference, which can be called "over-subtracting."

$$\$19.60 \xrightarrow{-\$8} \$11.60 \xrightarrow{+35¢} \$11.95$$

Chapter 15 Money   Notes

Students should practice mental math but should also be comfortable solving any of these problems using the vertical algorithm when the problem is not easy to calculate mentally.

$$
\begin{array}{r} \$17.80 \\ +\ \$9.60 \\ \hline \$27.40 \end{array} \rightarrow \begin{array}{r} 1{,}780 \\ +\ \ 960 \\ \hline 2{,}740 \end{array} \rightarrow \$27.40
$$

$$
\begin{array}{r} \$19.60 \\ -\ \$7.65 \\ \hline \$11.95 \end{array} \rightarrow \begin{array}{r} 1{,}960 \\ -\ \ 765 \\ \hline 1{,}195 \end{array} \rightarrow \$11.95
$$

Note that as in Dimensions Math 2B, this chapter will use the informal term "dot." The term "decimal point" is not used yet. Students familiar with the term already, may continue to use it. The emphasis in this chapter is merely that the dot separates the dollars from the cents.

Chapter 15 Money

Materials

- 10-sided die
- Bag of random coins
- Blank recording sheet
- Newspaper circulars
- Play coins and bills
- Whiteboards

Blackline Masters

- Checkbook Ledger
- Number Bond Template
- Pico, Fermi, Bagel

Activities

Activities included in this chapter are designed to provide practice with counting, adding, and subtracting money. They can be used after students complete the **Do** questions, or any time additional practice is needed.

Chapter Opener

Objective
- Investigate making change.

Materials
- Play coins and bills

The Chapter Opener is an opportunity for students to review changing dollars and cents to cents, and the notation for money.

They will also practice making change using strategies they have learned with numbers to 100.

Activities

▲ Trading Up

Materials: 10-sided die, play coins and bills

Review the value of coins. Students start with one quarter. On her turn, a player rolls the die and adds that amount of pennies to her start money.

If the amount added makes up another coin, she trades her change for the new coin.

The first player to get four quarters is the winner.

Sample play:

Player One rolls a 4. He adds 4 pennies to his quarter for 29 cents.

Player Two rolls a 6 and adds 6¢ to her money. She now has a quarter, a nickel, and a penny. (31¢)

Player One rolls a 6. He adds and exchanges his pennies. He now has either a quarter and two nickels or a quarter and one dime. (35¢)

Player Two rolls a 10. She collects pennies and makes exchanges to have a quarter, a dime, a nickel, and a penny. (41¢)

▲ Money War

Materials: Bag of random coins

Players take turns reaching into the bag and pulling out a handful of coins.

Players total the coins in each grab. The player with the highest value of coins scores a point.

The winner is the player with the most points after several rounds.

Lesson 1 Dollars and Cents

Objective

- Count and convert amounts of money within $100 from dollars and cents to cents, and from cents to dollars and cents.

Lesson Materials

- Play coins and bills

Think

Pose the **Think** problem and provide students with play coins and bills if needed.

Discuss how students counted the money.

Possible student solutions:

- Add the bills: $10 + $5 + $1 = $16
- Count on from the coins with the greatest value to coins with the least value.
- Group the coins first to make $1.

Learn

Have students compare their solutions from **Think** with the one shown in the textbook.

Emma and Mei are working together to figure out how much money Alex saved. Emma adds the bills and Mei totals the coins separately. They then add their two values together to get a total of $17.12.

Discuss Sofia's comment regarding the dot used to separate dollars and cents. When we read or write an amount in dollars and cents, we write the dollars first, then the cents after the "dot."

Dion points out that since $1 is equal to 100 cents, $17.00 is equal to 1,700¢.

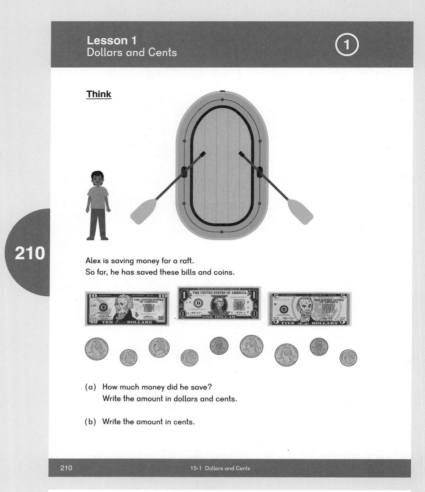

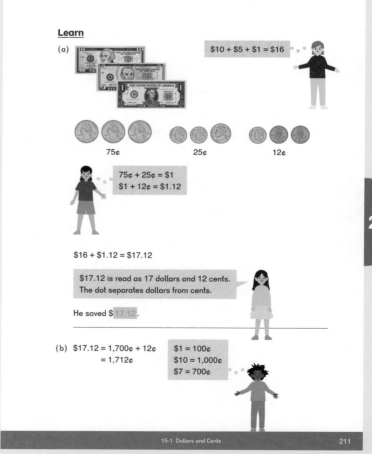

Do

❶—❷ Discuss the ways that the amounts are presented in dollars and cents and number bond format.

Mei and Dion think about the number bonds to help students think about how to convert the amounts.

❸ Have students share how they found the value of the coins. (I counted them all, I combined the quarters and dimes to make $1, etc.)

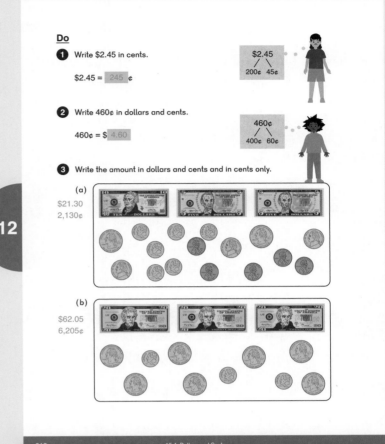

④ Alex reminds students that the dollars and cents can be separated, and then the dollars converted to cents.

⑤ Sofia knows that 100 cents = $1.00. She converts her cents to as many dollars as she can, with 43 cents left over.

(e) Remind students that they can think of 3¢ as 0 dollars and 3 pennies, so it is written as $0.03.

Activity

▲ Stumper

Materials: Play coins, Number Bond Template (BLM), play $1 bill.

Students play with partners. Using a Number Bond Template (BLM), place the $1 bill in the "whole."

Partners take turns being the Stumper and the Solver.

The Stumper puts some coins in one part of the number bond.

The Solver finds how much more money needed to make $1 and puts coins equaling that amount in the other part of the number bond.

If the Solver finds the correct value of coins for the missing part, he gets one point.

If he makes an error, the Stumper wins the point.

The winner is the player with the most points after the allotted time is up.

★ Instead of using coins, the Stumper and Solver write amounts in the number bond parts.

Exercise 1 • page 199

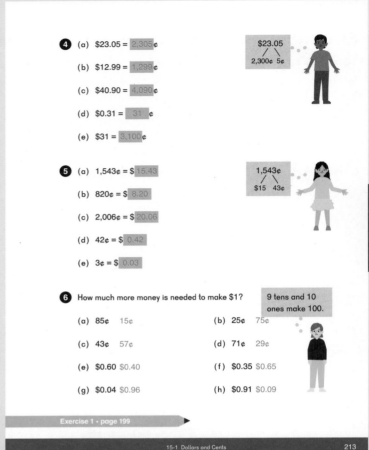

Lesson 2 Making $10

Objectives
- Add amounts that make $10.
- Subtract amounts from $10.

Think

Pose the **Think** problem and have students try to solve the problem independently.

Have students share their strategies for finding how much money Sofia needs to buy the bicycle pump.

Learn

Have students compare their solutions from **Think** with the ones shown in the textbook.

Alex knows:

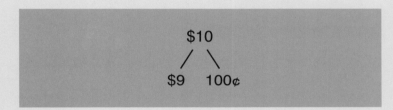

He adds 35¢ to 65¢ to make 100¢ or $1.
He adds $2 to the $7 to get $9.

$9 + $1 = $10

Dion uses a number bond to split 1,000¢ to make the next dollar first. He makes $8 by adding 35¢ to $7.65. He then adds $2 to get to $10.

Note that while both Alex and Dion are adding amounts on to $7.65, students may also subtract:

100¢ − 65¢ = 35¢
$9 − $7 = $2

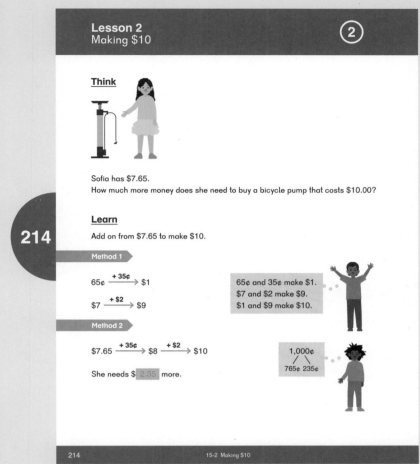

© 2017 Singapore Math Inc. Teacher's Guide 3B Chapter 15 245

Do

These problems (and additional ones) could be recreated on index cards as shopping tags and students could search to find the complementary amount to $10. Save the cards for future lessons.

Activity

▲ **Stumper to $10**

Students play with partners. Partners take turns being the Stumper and the Solver.

The Stumper writes an amount of money less than $10.

The Solver finds how much more money makes $10.

Exercise 2 • page 202

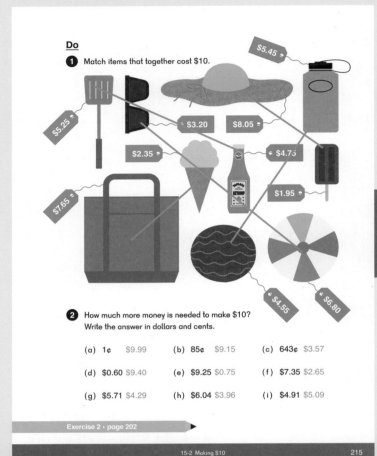

Lesson 3 Adding Money

Objective

- Add money amounts within $100.

Think

Pose the **Think** problem and have students try to solve the problem independently.

Discuss student strategies for solving the problem.

Learn

Have students compare their solutions from **Think** with the one shown in the textbook.

Mei adds the dollars first, then the cents, using a number bond to help her regroup to make the next dollar.

Emma uses a number bond to split $9.80. She adds 60 cents to $29.40 to make the next whole dollar amount, $30, then adds on the rest.

Sofia knows that adding $10 and subtracting 20¢ is the same as adding $9.80, and that it is easy to add $10.

Method 4 uses the vertical algorithm for money the way it is used for whole numbers. Students convert from dollars and cents, to all cents and then add just the cents. Dion notices that we always add dollars to dollars and cents to cents when we line the quantities up vertically. The dot stays in the same place.

Just like whole numbers, we are regrouping 100 cents as $1 and every ten $1 bills as $10.

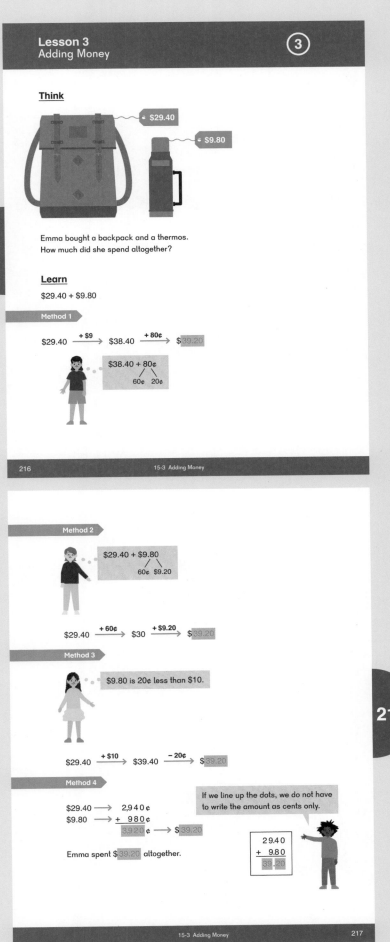

© 2017 Singapore Math Inc. Teacher's Guide 3B Chapter 15 247

Do

① Before solving the problems, have students discuss the patterns they notice in problems (a), (b), and (c).

(a) Students first add cents to make a whole dollar, then add more dollars.

(b)–(c) Students make the next dollar first.

② Each problem demonstrates a different strategy. Have students discuss why the computations are easily done using the corresponding method the friend uses.

For example, "Why is ② (c) appropriate for over-adding where ② (a) may not be?"

(d) reminds students that sometimes it is easier to compute the addition directly than to search for an appropriate strategy.

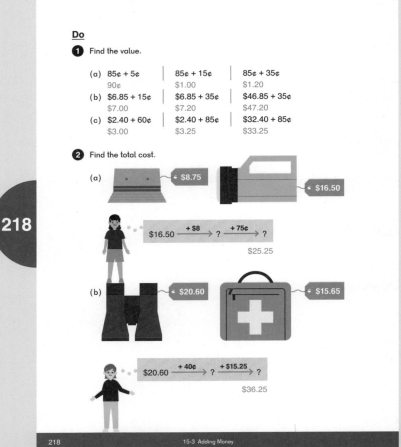

3 Students can use any of the methods they have learned. After they have worked the problems, have students share why they used their chosen methods.

Activity

▲ Shopping

Materials: Newspaper circulars

Provide students with different circulars and have them find the cost of purchasing two items. Use department or electronics store ad circulars to ensure there are prices between $10 and $100. Use grocery store ad circulars for students who need additional practice with sums less than $20.

This is a good activity to develop the mental math strategies from Dimensions Math 3A Chapter 2: Addition and Subtraction — Part 1 of adding numbers close to 100 and 1,000, as many items end in 98 or 99 cents.

For example:

$9.98 + $19.99
Students may think, "$10 + $20 − $0.03."

Exercise 3 • page 204

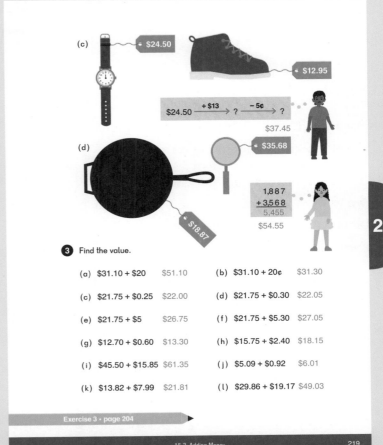

Lesson 4 Subtracting Money

Objective

- Subtract money amounts within $100.

Think

Pose the **Think** problem and have students try to solve the problem independently.

Students may immediately note that they can use the vertical algorithm method discussed in previous lessons. If so, challenge them to find a mental method as well.

Discuss student strategies for solving the problem.

Learn

Have students compare their solutions from **Think** with the ones shown in the textbook.

Emma splits 85¢, so she can subtract the dollars first, then the cents.

Sofia subtracts 50¢ first, to get to 18 whole dollars, then subtracts the remaining $6.35 from $18.

Dion knows that subtracting $6.85 is the same as subtracting $7 and then adding back 15¢.

Alex uses the vertical algorithm for money the way it is used for whole numbers. Students can think about how they convert $18.50 and $6.85 to cents and then subtract just cents.

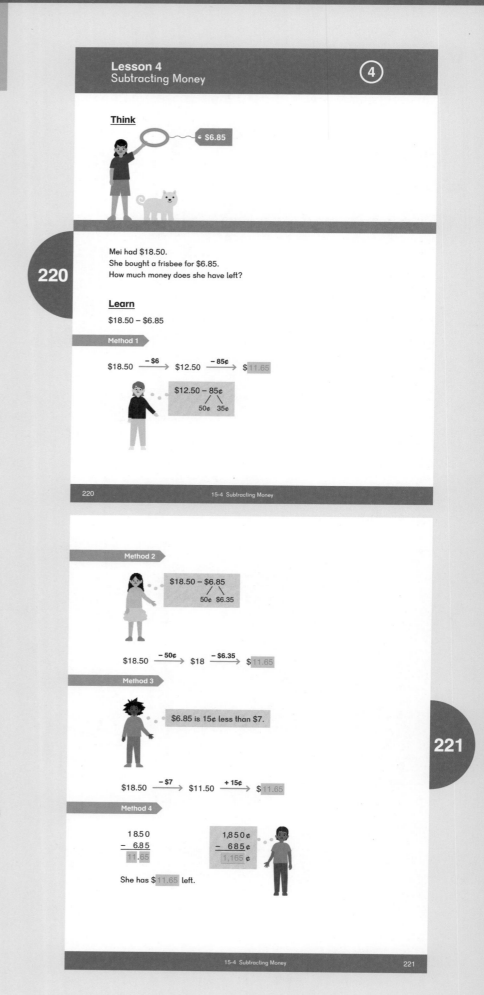

Do

1 (a) 40¢ and 60¢ make a dollar, so 40¢ and $7.60 make $8.

(b) Remind students who are struggling that:

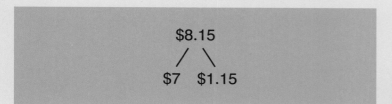

2 Each problem demonstrates a different strategy. Have students discuss why each computation is easily done using the corresponding method the friend uses.

For example, "Why is **2** (e) easier to do by over-subtracting where **2** (c) may not be?"

3 Students can use any of the methods they have learned. After they have worked the problems, have students share why they used their chosen methods.

Activity

▲ Allowance

Materials: Newspaper circulars

Provide students with a set amount of money within $100 as an allowance. Use either newspaper circulars or index cards with item costs. Students subtract the cost of the item they would like to purchase from their allowance and see how much money they have left. Have students find a second item and subtract that from their remaining money.

★ See who can buy the most unique items with their total amount of money.

Exercise 4 • page 207

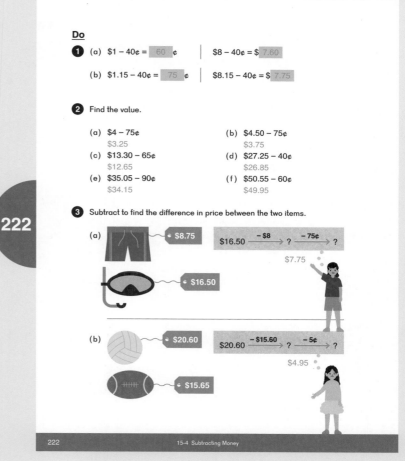

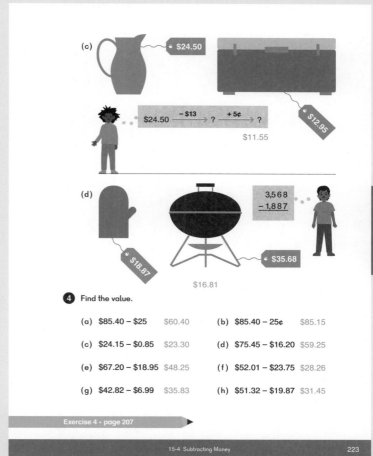

Lesson 5 Word Problems

Objective

- Solve word problems involving adding and subtracting money.

Think

Pose the **Think** problem and have students try to solve the problem independently.

Have students draw bar models to set up the solution.

Discuss student solutions.

Learn

Have students compare their solutions from **Think** with the one shown in the textbook. The **Learn** example solves the problem by adding the cost of the bed float and flamingo float first, then subtracting that total from $40.

Dion chooses to subtract the cost of the bed float from $40 first, then he subtracts the cost of the flamingo float from his remaining money ($25.10).

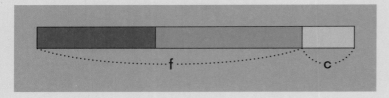

Students can also solve the problem using equations, with a letter standing for the unknown quantity:

f = cost of the two floats
c = amount received in change

f = $14.90 + $18.50
c = $40 − f

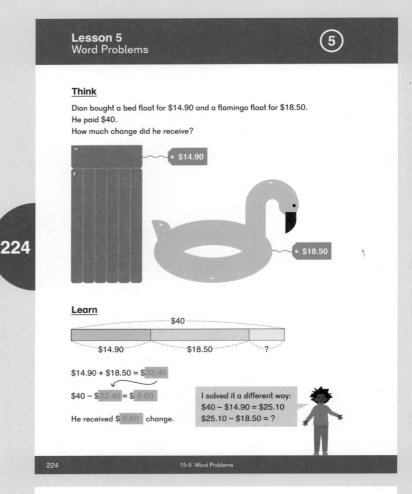

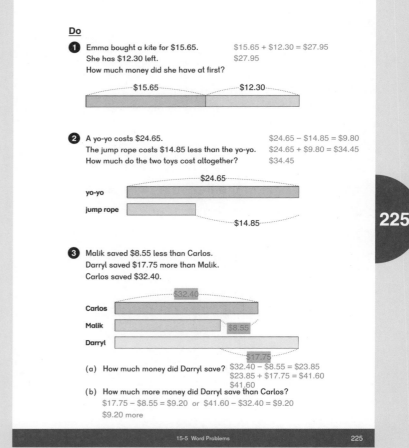

252   Teacher's Guide 3B Chapter 15   © 2017 Singapore Math Inc.

Do

1 – **4** Discuss the models. Ask students:

- What do we need to find? Where do you think a question mark might go on the model?
- Are we comparing items, or are we trying to find a part or a whole?
- What do we know?
- Do we have enough information to find what is unknown?

2 Ensure that students understand the amount given is not the amount of the jump rope, but the amount by which the jump rope costs less than the yo-yo.

3 Ask students:

- "What do you need to find?" (How much more Darryl saved than Carlos.)
- "What information is given?" (How much Carlos saved.)
- "What do you need to find to determine how much Darryl saved?" (Find how much Malik saved first).

4 – **6** Continue asking prompting questions as needed.

Students can also solve the following problems using equations, with a letter standing for the unknown quantity:

2 j = cost of the jump rope
t = total cost of the two toys

j = $24.65 − $14.85
t = j + $24.65

3 m = amount Malik saved
d = amount Darryl saved

m = $32.40 − $8.55
d = m + $17.75

4 a = amount three people contribute
b = difference between contributed amount and cost of canoe

a = $32 × 3
b = a − $89.45

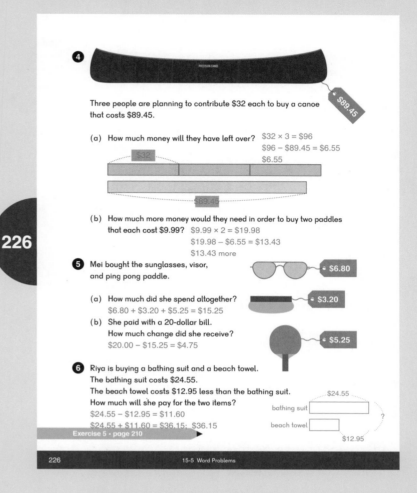

6 t = cost of the beach towel
u = cost of the two items

t = $24.55 − $12.95
u = t + $24.55

Exercise 5 • page 210

Lesson 6 Practice

Objective

- Practice skills from the chapter.

After students complete the **Practice** in the textbook, have them continue adding and subtracting money by playing games from the chapter.

5 — 10 These problems can elicit much discussion. Students may draw a model for **5 — 7**, if needed. Most should be able to solve the problem without the model.

8 — 10 Students may need to draw a bar model to solve these problems as they are more challenging.

Activity

▲ Classroom Economy

Materials: Checkbook Ledger (BLM)

Setting up a classroom economy for a week or even a month can be a great way to practice adding and subtracting money. Provide each student with a Checkbook Ledger (BLM) and a weekly allowance. Have students earn money by completing homework or tasks in the classroom. Have them pay for privileges like borrowing a pencil or renting their chairs.

Students record each deposit or payment in their ledgers.

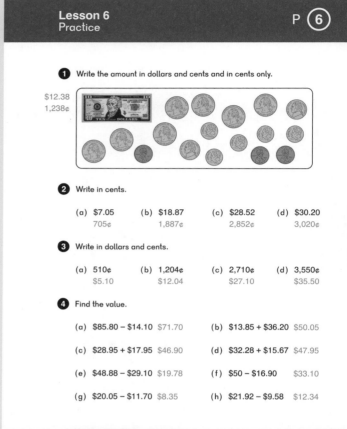

⑧ – ⑩ Students can also solve the problem using equations, with a letter standing for the unknown quantity:

⑧ t = cost of pail and shovel
c = amount received in change

t = $5.45 + $ 2.35
c = $10 – t

⑩ a = cost of pants and shoes
s = cost of 3 t-shirts
t = cost of 1 t-shirt

a = $19.85 + $24.15
s = $83 – a
t = s ÷ 3

Brain Works

★ Pico, Fermi, Bagel

Materials: Blank recording sheet or Pico, Fermi, Bagel (BLM)

Player One thinks of or writes down a three-digit number. Player Two tries to guess the number. Player One scores each round's guess with a Pico, Fermi, Bagel according to the placement of the numbers.

Rules:
Pico – Correct number, wrong place
Fermi – Correct number and place
Bagel – No correct numbers

Example:
Player One chooses the number 587.

On his first guess, Player Two chooses 123. Since none of those numbers are in 587, Player One writes B, B, B.

On his second guess, Player Two chooses 456. The number 5 is in 587, but not in the correct place as guessed. Player One writes B,P,B; and Player Two knows that either the first or third digit is a 5, and that the other two numbers are either 0, 7, 8, or 9.

Play continues until Player Two chooses the right number, then they trade roles.

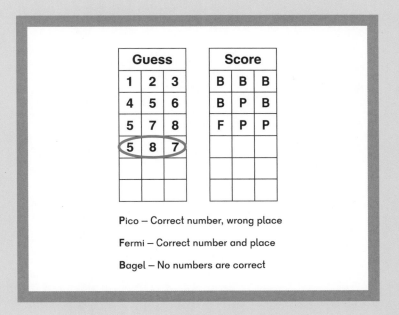

Exercise 6 • page 213

Review 4

Objective

- Cumulative review of content from Chapter 1 through Chapter 15.

Use the cumulative reviews as necessary to practice and reinforce content and skills from 3A and 3B.

Review 4 generally covers numbers and data content.

Review 5 generally covers geometry and measurement content.

Students should have mastered the content from Dimensions Math 3A and 3B to successfully start Dimensions Math 4A. Use games and activities from all prior chapters as needed to solidify facts and skills.

Review 4 R ④

① (a) 763 ÷ 9 is 84 with a remainder of 7.

(b) 59 × 6 = 354

② What is the greatest even number that can be formed using the digits 7, 8, 4, and 3?
8,734

③

```
0       P    Q       R    1       S
```

(a) What fractions are indicated by P, Q, and R on this number line? Give your answer in simplest form.
P: $\frac{1}{3}$, Q: $\frac{1}{2}$, R: $\frac{5}{6}$

(b) Give an equivalent fraction for R with a denominator of 12.
$\frac{10}{12}$

(c) S is how many thirds?
4

④ (a) Which of the following are greater than $\frac{1}{2}$?

(b) Put the fractions in order from least to greatest.
$\frac{3}{10}, \frac{3}{8}, \frac{3}{5}, \frac{5}{6}, \frac{8}{9}, \frac{9}{8}$

229

---

⑤ This table and graph show the number of people who visited City Zoo during one week.

Day	Number
Sunday	442
Monday	308
Tuesday	172
Wednesday	489
Thursday	225
Friday	371
Saturday	567

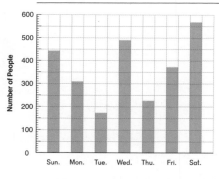

(a) The scale on the left is numbered in increments of 100.

(b) Each tick mark shows an increment of 10.

(c) Each square on the graph is for an increment of 50.

230

---

256   Teacher's Guide 3B Chapter 15   © 2017 Singapore Math Inc.

6 (c) Students can also solve the problem using equations, with a letter standing for the unknown quantity:

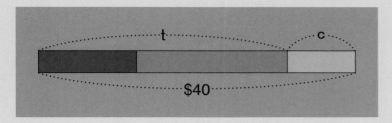

t = cost of toy panda and t-shirt
c = amount received in change

t = $12.40 + $18.95
c = $40 − t

Exercise 7 • page 217

(d) Which day of the week was the most popular for visiting the zoo?
Saturday
(e) Which day of the week was the least popular for visiting the zoo?
Tuesday
(f) How many more people visited the zoo on Saturday than on Sunday?
567 − 442 = 125; 125 more people
(g) Estimate how many people visited the zoo during the weekdays by rounding to the nearest 100.   300 + 200 + 500 + 200 + 400 = 1,600
About 1,600 people
(h) On the weekend, tickets are $8 each.
How much money did the zoo receive from tickets on Saturday?
567 × $8 = $4,536; $4,536
(i) On Wednesdays, tickets are $5 each.
How much money did the zoo receive from tickets on Wednesday?
489 × $5 = $2,445; $2,445
(j) For the rest of the week, tickets are $6 each.
How much less money did the zoo receive from tickets on Tuesday than on Monday?   or   308 × $6 = $1,848; 172 × $6 = $1,032
308 − 172 = 136; 136 × $6 = $816 $1,848 − $1,032 = $816; $816 less
(k) During the first half-hour on Tuesday, the total ticket sales was $294.
How many people bought tickets during the half-hour? $294 ÷ 6 = 49
49 people
(l) During the first half-hour on Saturday morning, the total ticket sales was $928 from selling tickets at $8 each.
During the next half-hour, the total ticket sales was $472.
How many more people bought tickets during the first half-hour than the second half-hour? $928 − $472 = $456
$456 ÷ $8 = 57
57 more people

231

---

6 (a) On weekdays, the zoo opens at 10:00 a.m. and closes at 4:30 p.m.
How long is it open each weekday?
6 h 30 min
(b) On weekends, the zoo opens 1 h 30 min earlier, and closes at 6:00 p.m.
How long is it open each weekend day?
9 h 30 min
(c) At the zoo, Avery bought a toy panda for $12.40 and a t-shirt for $18.95.
She paid with two twenty-dollar bills. $12.40 + $18.95 = $31.35
How much change did she receive? $40 − $31.35 = $8.65
$8.65
(d) Avery left the zoo at 3:45 p.m.
She was there for five and a half hours.
When did she arrive at the zoo?
10:15 a.m.

7 One day, a gorilla ate $\frac{5}{8}$ kg of broccoli and $\frac{3}{8}$ kg of kale.

(a) How many kilograms of both vegetables did it eat?
$\frac{5}{8}$ kg + $\frac{3}{8}$ kg = $\frac{8}{8}$ kg = 1 kg; 1 kg
(b) How many more kilograms of broccoli did it eat than kale?
$\frac{5}{8}$ kg − $\frac{3}{8}$ kg = $\frac{2}{8}$ kg = $\frac{1}{4}$ kg; $\frac{1}{4}$ kg

8 An elephant at the zoo drank 95 L of water one day.
A giraffe drank 27 L 500 mL that day.
How much more did the elephant drink than the giraffe?
95 L − 27 L 500 mL = 67 L 500 mL

Exercise 7 • page 217

Review 5

Objective

- Cumulative review of content from Chapter 1 through Chapter 15.

Use the cumulative reviews as necessary to practice and reinforce content and skills from 3A and 3B.

7 Students can also solve the problem using equations, with a letter standing for the unknown quantity:

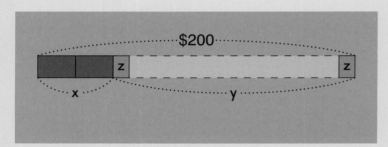

x = cost of two bats
y = amount left to spend on baseballs
z = number of baseballs the coach can buy

x = $24 × 2
y = $200 − x
z = y ÷ $4

Review 5 R 5

1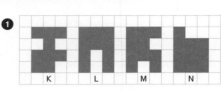

K Perimeter: 18 units
K Area: 9 sq. units
L Perimeter: 18 units
L Area: 10 sq. units
M Perimeter: 18 units
M Area: 10 sq. units
N Perimeter: 14 units
N Area: 10 sq. units

(a) Which figure has a different perimeter from the other three figures?
Figure N
(b) Which figure has a different area from the other three figures?
Figure K

2 This wall decoration is made from eight triangles.
Each triangle has 3 equal sides.
The perimeter of each triangle is 72 cm.
72 cm ÷ 3 = 24 cm; One side of each triangle is 24 cm.
(a) What is the perimeter of the wall decoration?
Give your answer in meters and centimeters.
24 cm × 8 = 192 cm; 1 m 92 cm
(b) Is the wall decoration a rhombus?
Explain why or why not.
Yes, it has 4 equal sides.
(c) What fraction of it is colored green?
Give your answer in simplest form.
$\frac{1}{4}$
(d) Which of the labeled angles are larger than a right angle?
Angles s and t.
(e) Which of the labeled angles are smaller than a right angle?
Angles r and u.

233

3 It took Cooper 15 minutes to fall asleep after he went to bed. He then slept soundly for 4 h 55 min, then was awake for 15 minutes, and then slept again for 3 h 20 min.

(a) How long did he sleep altogether?
4 h 55 min + 3 h 20 min = 8 h 15 min
(b) He went to bed at 10:30 p.m. and got up as soon as he woke up. What time did he get up?
8 h 15 min + 15 min + 15 min = 8 h 45 min
7:15 a.m.

4 Tina got back from a hike at 2:30 p.m.
She was gone for 4 h 15 min.
What time did she leave?
10:15 a.m.

5 A rectangular field has a length of 120 m and a width of 85 m.
Kawai ran around the field 6 times.
How far did he run? 120 m + 85 m = 205 m
Give your answer in kilometers and meters. 205 m × 2 = 410 m = 1 lap
410 m × 6 = 2,460 m = 2 km 460 m

6 The radius of this circle is 10 cm.
The two shorter sides of the triangle
are 12 cm and 16 cm long.
What is the perimeter of this triangle?
20 cm + 12 cm + 16 cm = 48 cm
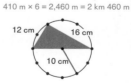

7 Bats cost $24 each and baseballs cost $4 each.
A coach has $200.
After he buys 2 bats, how many baseballs can he buy?
2 bats: $24 × 2 = $48
$200 − $48 = $152
$152 ÷ $4 = 38
He can buy 38 baseballs.

234

Exercise 8 • page 222

8 A line drawn from which point, B, C, or D, to point A, will form a right angle with this blue line? Point C

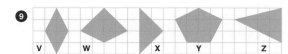

9

(a) Which figures are quadrilaterals?
Figures V and W.
(b) Which triangle has two equal angles?
Figure X.
(c) Which figures have a right angle?
Figures W and X.
(d) Which figure is a rhombus?
Figure V.
(e) Find the area of each figure.
V = 4 sq. units   X = 4 sq. units   Z = 6 sq. units
W = 6 sq. units   Y = 8 sq. units

10 The diameter of the largest circle is 112 cm.
What are the radii and diameters of the smaller circles?
112 cm ÷ 2 = 56 cm   112 cm ÷ 4 = 28 cm   or   56 cm ÷ 2 = 28 cm
Diameter = 56 cm   Radius = 28 cm

11 When the time is 12:15, do the hands on a clock form a right angle?
No, it is a less than a right angle since the hour hand points a little past the 12.

12 Carpeting costs $6 per square meter to install.
The carpet itself costs $2 per square meter.
How much will it cost to carpet this room?

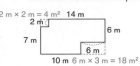

10 m   6 m × 3 m = 18 m²

16 m × 9 m = 144 m²
Area: 144 m² − 18 m² − 4 m² = 122 m²
$8 × 122 = $976
$976

13 This design was made from squares each with sides of 6 cm.

(a) What fraction of the design is blue?
Give your answer in simplest form.
$\frac{2}{3}$
(b) What is the perimeter of the design?
96 cm
(c) What is the area of the design?
6 cm × 6 cm = 36 cm²; 36 cm² × 9 = 324 cm²
324 cm²

14 The length of this glass and wire decoration is 45 centimeters.
The wire used to make the frame around each square piece of glass was cut from some wire that was 3 meters long.
Each larger square is twice as long as the smaller square.
5 units = 45 cm; 1 unit = 45 cm ÷ 5 = 9 cm
(a) What is the length of the leftover piece of wire?
20 × 9 cm = 180 cm; 3 m − 1 m 80 cm = 1 m 20 cm
(b) What is the total area of the glass used in the decoration, in square centimeters?
Area of smaller square: 9 cm × 9 cm = 81 cm²
Total area = 81 cm² × 9 = 729 cm²

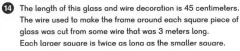

Exercise 1 • pages 199–201

Chapter 8 Multiplying and Dividing with 6, 7, 8, and 9

Exercise 1

Basics

1. (a) 6 × 6 is 6 more than [5] × 6.

    6 × 6 = [36]

    (b) 9 × 6 is 6 less than [10] × 6.

    9 × 6 = [54]

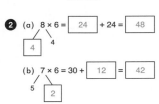

2. (a) 8 × 6 = [24] + 24 = [48]

    with branches 4, 4

    (b) 7 × 6 = 30 + [12] = [42]

    with branches 5, [2]

3. 6 × 9 = 9 × [6] = [54]

4. (a) [9] × 6 = 54   54 ÷ 6 = [9]

    (b) [6] × 6 = 36   36 ÷ 6 = [6]

5. 42 ÷ 6 = [7]   42 + [3] = 45

    45 ÷ 6 is [7] with a remainder of [3].

8-1 The Multiplication Table of 6

Exercise 2 • pages 202–203

Exercise 3 • pages 204–206

Chapter 8 Multiplying and Dividing with 6, 7, 8, and 9

Exercise 1

Basics

1. (a) 6 × 6 is 6 more than [5] × 6.

    6 × 6 = [36]

    (b) 9 × 6 is 6 less than [10] × 6.

    9 × 6 = [54]

2. (a) 8 × 6 = [24] + 24 = [48]

    [4] 4

    (b) 7 × 6 = 30 + [12] = [42]

    5 [2]

3. 6 × 9 = 9 × [6] = [54]

4. (a) [9] × 6 = 54   54 ÷ 6 = [9]

   (b) [6] × 6 = 36   36 ÷ 6 = [6]

5. 42 ÷ 6 = [7]   42 + [3] = 45

   45 ÷ 6 is [7] with a remainder of [3].

8-1 The Multiplication Table of 6

Exercise 4 • pages 207–209

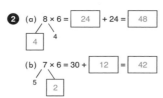
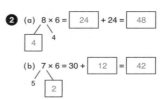

Exercise 5 • pages 210–212

Chapter 8 Multiplying and Dividing with 6, 7, 8, and 9
Exercise 1
Basics

1. (a) 6 × 6 is 6 more than [5] × 6.
   6 × 6 = [36]
   (b) 9 × 6 is 6 less than [10] × 6.
   9 × 6 = [54]

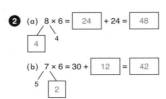

2. (a) 8 × 6 = [24] + 24 = [48]
           / \
          4   4
   (b) 7 × 6 = 30 + [12] = [42]
           / \
          5   [2]

3. 6 × 9 = 9 × [6] = [54]

4. (a) [9] × 6 = 54 | 54 ÷ 6 = [9]
   (b) [6] × 6 = 36 | 36 ÷ 6 = [6]

5. 42 ÷ 6 = [7] | 42 + [3] = 45
   45 ÷ 6 is [7] with a remainder of [3].

8-1 The Multiplication Table of 6   1

Chapter 8 Multiplying and Dividing with 6, 7, 8, and 9
Exercise 1
Basics

1. (a) 6 × 6 is 6 more than [5] × 6.
   6 × 6 = [36]
   (b) 9 × 6 is 6 less than [10] × 6.
   9 × 6 = [54]

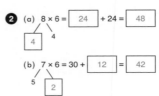

2. (a) 8 × 6 = [24] + 24 = [48]
           / \
          4   4
   (b) 7 × 6 = 30 + [12] = [42]
           / \
          5   [2]

3. 6 × 9 = 9 × [6] = [54]

4. (a) [9] × 6 = 54 | 54 ÷ 6 = [9]
   (b) [6] × 6 = 36 | 36 ÷ 6 = [6]

5. 42 ÷ 6 = [7] | 42 + [3] = 45
   45 ÷ 6 is [7] with a remainder of [3].

8-1 The Multiplication Table of 6   1

Chapter 8 Multiplying and Dividing with 6, 7, 8, and 9
Exercise 1
Basics

1. (a) 6 × 6 is 6 more than [5] × 6.
   6 × 6 = [36]
   (b) 9 × 6 is 6 less than [10] × 6.
   9 × 6 = [54]

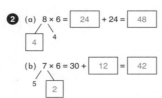

2. (a) 8 × 6 = [24] + 24 = [48]
           / \
          4   4
   (b) 7 × 6 = 30 + [12] = [42]
           / \
          5   [2]

3. 6 × 9 = 9 × [6] = [54]

4. (a) [9] × 6 = 54 | 54 ÷ 6 = [9]
   (b) [6] × 6 = 36 | 36 ÷ 6 = [6]

5. 42 ÷ 6 = [7] | 42 + [3] = 45
   45 ÷ 6 is [7] with a remainder of [3].

8-1 The Multiplication Table of 6   1

Exercise 6 • pages 213–216

Chapter 8 Multiplying and Dividing with 6, 7, 8, and 9
Exercise 1
Basics

1. (a) 6 × 6 is 6 more than [5] × 6.
   6 × 6 = [36]
   (b) 9 × 6 is 6 less than [10] × 6.
   9 × 6 = [54]

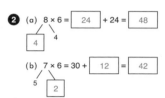

2. (a) 8 × 6 = [24] + 24 = [48]
         /\
        4 4
   (b) 7 × 6 = 30 + [12] = [42]
         /\
        5 2

3. 6 × 9 = 9 × [6] = [54]

4. (a) [9] × 6 = 54   54 ÷ 6 = [9]
   (b) [6] × 6 = 36   36 ÷ 6 = [6]

5. 42 ÷ 6 = [7]   42 + [3] = 45
   45 ÷ 6 is [7] with a remainder of [3].

8-1 The Multiplication Table of 6   1

Chapter 8 Multiplying and Dividing with 6, 7, 8, and 9
Exercise 1
Basics

1. (a) 6 × 6 is 6 more than [5] × 6.
   6 × 6 = [36]
   (b) 9 × 6 is 6 less than [10] × 6.
   9 × 6 = [54]

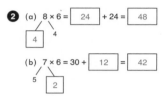

2. (a) 8 × 6 = [24] + 24 = [48]
         /\
        4 4
   (b) 7 × 6 = 30 + [12] = [42]
         /\
        5 2

3. 6 × 9 = 9 × [6] = [54]

4. (a) [9] × 6 = 54   54 ÷ 6 = [9]
   (b) [6] × 6 = 36   36 ÷ 6 = [6]

5. 42 ÷ 6 = [7]   42 + [3] = 45
   45 ÷ 6 is [7] with a remainder of [3].

8-1 The Multiplication Table of 6   1

© 2017 Singapore Math Inc.   Teacher's Guide 3B Chapter 15   265

Exercise 7 • pages 217–221

Chapter 8 Multiplying and Dividing with 6, 7, 8, and 9

Exercise 1

Basics

1. (a) 6 × 6 is 6 more than [5] × 6.

    6 × 6 = [36]

    (b) 9 × 6 is 6 less than [10] × 6.

    9 × 6 = [54]

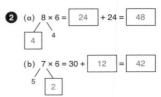

2. (a) 8 × 6 = [24] + 24 = [48]

    4 ⌄ 4

    (b) 7 × 6 = 30 + [12] = [42]

    5 ⌄ 2

3. 6 × 9 = 9 × [6] = [54]

4. (a) [9] × 6 = 54 | 54 ÷ 6 = [9]

    (b) [6] × 6 = 36 | 36 ÷ 6 = [6]

5. 42 ÷ 6 = [7] | 42 + [3] = 45

    45 ÷ 6 is [7] with a remainder of [3].

8-1 The Multiplication Table of 6

Chapter 8 Multiplying and Dividing with 6, 7, 8, and 9

Exercise 1

Basics

1. (a) 6 × 6 is 6 more than [5] × 6.

    6 × 6 = [36]

    (b) 9 × 6 is 6 less than [10] × 6.

    9 × 6 = [54]

2. (a) 8 × 6 = [24] + 24 = [48]

    4 ⌄ 4

    (b) 7 × 6 = 30 + [12] = [42]

    5 ⌄ 2

3. 6 × 9 = 9 × [6] = [54]

4. (a) [9] × 6 = 54 | 54 ÷ 6 = [9]

    (b) [6] × 6 = 36 | 36 ÷ 6 = [6]

5. 42 ÷ 6 = [7] | 42 + [3] = 45

    45 ÷ 6 is [7] with a remainder of [3].

8-1 The Multiplication Table of 6

Chapter 8 Multiplying and Dividing with 6, 7, 8, and 9

Exercise 1

Basics

1. (a) 6 × 6 is 6 more than [5] × 6.

    6 × 6 = [36]

    (b) 9 × 6 is 6 less than [10] × 6.

    9 × 6 = [54]

2. (a) 8 × 6 = [24] + 24 = [48]

    4 ⌄ 4

    (b) 7 × 6 = 30 + [12] = [42]

    5 ⌄ 2

3. 6 × 9 = 9 × [6] = [54]

4. (a) [9] × 6 = 54 | 54 ÷ 6 = [9]

    (b) [6] × 6 = 36 | 36 ÷ 6 = [6]

5. 42 ÷ 6 = [7] | 42 + [3] = 45

    45 ÷ 6 is [7] with a remainder of [3].

8-1 The Multiplication Table of 6

Chapter 8 Multiplying and Dividing with 6, 7, 8, and 9

Exercise 1

Basics

1. (a) 6 × 6 is 6 more than [5] × 6.

    6 × 6 = [36]

    (b) 9 × 6 is 6 less than [10] × 6.

    9 × 6 = [54]

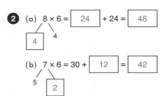

2. (a) 8 × 6 = [24] + 24 = [48]

    4 ⌄ 4

    (b) 7 × 6 = 30 + [12] = [42]

    5 ⌄ 2

3. 6 × 9 = 9 × [6] = [54]

4. (a) [9] × 6 = 54 | 54 ÷ 6 = [9]

    (b) [6] × 6 = 36 | 36 ÷ 6 = [6]

5. 42 ÷ 6 = [7] | 42 + [3] = 45

    45 ÷ 6 is [7] with a remainder of [3].

8-1 The Multiplication Table of 6

Exercise 8 • pages 222–226

Chapter 8 Multiplying and Dividing with 6, 7, 8, and 9

Exercise 1

Basics

1. (a) 6 × 6 is 6 more than [5] × 6.

   6 × 6 = [36]

   (b) 9 × 6 is 6 less than [10] × 6.

   9 × 6 = [54]

2. (a) 8 × 6 = [24] + 24 = [48]

      [4] 4

   (b) 7 × 6 = 30 + [12] = [42]

      5 [2]

3. 6 × 9 = 9 × [6] = [54]

4. (a) [9] × 6 = 54   54 ÷ 6 = [9]

   (b) [6] × 6 = 36   36 ÷ 6 = [6]

5. 42 ÷ 6 = [7]   42 + [3] = 45

   45 ÷ 6 is [7] with a remainder of [3].

8-1 The Multiplication Table of 6

Blackline Masters for 3B

All Blackline Masters used in the guide can be downloaded from dimensionsmath.com.
This lists BLMs used in the **Think** and **Learn** sections.
BLMs used in **Activities** are included in the Materials list within each chapter.

Area of Composite Figures	Chapter 13: Lesson 4
Array Dot Cards — 6	Chapter 8: Lesson 1
Array Dot Cards — 7	Chapter 8: Lesson 2
Centimeter Graph Paper	Chapter 12: Lesson 7, Chapter 13: Lesson 2, 4, 8
Centimeter Ruler	Chapter 12: Lesson 1
Circle	Chapter 12: Lesson 1
Circle Dot Paper	Chapter 12: Lesson 5
Comparing Fractions with Like Denominators	Chapter 9: Lesson 3
Fraction Chart	Chapter 10: Lesson 1
Match Time Cards	Chapter 14: Lesson 1
Multiplication Chart	Chapter 8: Chapter Opener
Multiplication Chart — 6	Chapter 8: Lesson 1
Multiplication Chart — 7	Chapter 8: Lesson 2
Multiplication Chart — 8	Chapter 8: Lesson 6
Multiplication Chart — 9	Chapter 8: Lesson 7
Paint Problem	Chapter 10: Lesson 7, Lesson 8
Square and Rectangle	Chapter 13: Lesson 1
Tangram	Chapter 12: Chapter Opener, Lesson 6
Timeline	Chapter 14: Lesson 4, Lesson 5, Lesson 6
Triangle	Chapter 12: Lesson 5

Notes